PARADIGM SHIFT

How Expert Opinions Keep Changing on Life, the Universe, and Everything

Martin Cohen

ia

imprint-academic.com

Published in the UK by
Imprint Academic, PO Box 200, Exeter EX5 5YX, UK

Distributed in the USA by
Ingram Book Company,
One Ingram Blvd., La Vergne, TN 37086, USA

ISBN 9781845407940

A CIP catalogue record for this book is available from the
British Library and US Library of Congress

Also by Martin Cohen

Political Philosophy: from Plato to Mao

'This unusual book, hard to classify, is almost a handbook of the history of political philosophy... a very interesting retrospective... recommended for sophisticated undergraduates and above.'
—S.C. Schwarze, Cabrini College in *CHOICE Academic Reviews*

'...a broad sweep covered with surprising agility and clarity... The central advantages are undoubtably its lucidity, range and unorthodox approach.'
— *Times Higher Education Supplement*

The Doomsday Machine: The High Price of Nuclear Energy, the World's Most Dangerous Fuel

'A polemic on the evils of splitting the atom.'
— *New York Times*

101 Philosophy Problems

'Cohen continually delights or infuriates us with his irreverent opinions. He tells us, for example, that Kant reduced philosophy "to esoteric monologues of professionals" and that Aristotle "suffered from a particularly severe taxonomical disorder". Logic is irrelevant, a point he reinforces by not using it to clarify philosophical problems.'
— *Times Higher Education Supplement* (London)

'Tired of yet more introductions, anthologies and text books, publishers are beginning to wake up to the fact that, especially for the non-academic reader, what is needed in philosophy are different kinds of books that can engage the interest of the enthusiast... It has long been recognised that philosophy is, among other things, something that needs to be engaged in. You can't just read philosophy, you've got to actually do it. Given that, it's surprising how few introductions actually try and get their readers to join in.'
— *The Philosopher's Magazine*

'Without technical terms and incomprehensible superstructure, but with wit and irony, Cohen explains the basic concepts of philosophy and, in passing, the most famous thinkers of history.'
— *Der Spiegel*

101 Ethical Dilemmas

'...a chatty, jokey journey through philosophical dilemmas, ancient and modern... but the philosophy is the real thing.'
— *New Scientist*

'There are a lot of DIY philosophy texts around — not so much takin' it to the streets as takin' it to the dinner parties. This one isn't all that inspired but it is fun enough and if it actually gets people thinking about the issues, then it has

done its job. These dilemmas (from the Greek meaning 'two horns') cover a wide range of topics, and in addition there are 101 ethical discussions.
— *The Age* (Melbourne)

Wittgenstein's Beetle

'One of the fun things about philosophy as opposed to, say, chemistry is that to find things out you don't need to mess around with test tubes and Liebig condensers. Instead, you can sit back in your armchair, set up a laboratory in your own head and calmly observe the results of mixing x with y. This is the grand tradition of the "thought experiment", to which Cohen provides a zippy alphabetical guide.'
— *The Guardian*

'With its sense of history, *Wittgenstein's Beetle* provides the opportunity to consider which thought experiments last. Some are as surprising today as when they were first mooted. Plutarch rehearsed the one about falling under gravity to the centre of the earth, and then out again the other side: it produces the amazing result that the time taken to tumble between any two cities on the surface of the globe, via the straightest tunnel connecting them through the planet, would always be forty-three minutes.'
— *Times Literary Supplement*

Philosophical Tales

'In *Philosophical Tales*, Cohen presents what he calls an alternative history of philosophy that tells its story by interweaving brief biographies of the major philosophical players with brief discussions of their ideas, including many of their more frequently overlooked ideas. He attempts to show through a combination of humor and ironic irreverence that philosophy, or at least the lives of the great philosophers, is as filled with prejudices, jealousies, vanities, and ridiculous ideas, as it is with the pursuit of wisdom.

Through this unflattering tale of philosophy, Cohen aims to show that philosophers are ordinary people and that ordinary people are and should be philosophers. It is easy to suppose that Cohen's purpose is to cut to size or debunk the claims to intellectual superiority of great philosophers, much as Paul Johnson tried to do with regard to writers and thinkers in *Intellectuals*. Yet, if this is Cohen's purpose, it is not his only purpose. One of his important concerns is to raise some uncomfortable questions about philosophy and its great figures — questions to which historians of philosophy have not paid sufficient attention to.'
— *Teaching Philosophy*

Contents

Introduction

'I have opinions of my own — strong opinions — but I don't always agree with them.' — George W. Bush

Where do our opinions come from? The answer is more subtle than you might suppose. Experts, it turns out, are often following (or attempting to follow) other 'experts' — and the results can be disastrous. From the sub-prime disaster of 2007 that left national economies (and individuals' life savings) in tatters, to the approved medications that years later turn out to be not only useless but deadly dangerous, what seems to be solid, reliable information turns out to be wrong, often wildly so.

The idea behind this book is to treat a broad sweep of issues, ranging from public health to climate change and even high finance, as a series of 'case studies' collectively challenging the view so often put about of science and knowledge generally, as being a very sensible and reassuringly solid sort of affair. Each case study/story naturally tends to highlight one aspect of the philosophical quest for a theory of knowledge.

So, in the chapters that follow, I step gingerly through a wide swathe of modern life and conventional opinion, in an effort to highlight the illogicality, the inconsistencies, and the downright dishonesty of much of what we are repeatedly told is expert opinion, scholarly insight, settled fact. Bear with me, because it is only as that great philosophical fertilizer, doubt, is spread around that we really start to feel a need for the profound reflection that the world really demands of us. But the aim is not to advance a version of scepticism, ancient and modern, which seems to lead on to a further conclusion that we might as well give up serious thinking and do nothing, or act on 'whatever we happen to believe', be it by instinct or by custom or by 'accident'. Acknowledging uncertainty and complexity is, on the contrary, a positive step towards knowledge, just as Plato and Descartes insisted, so long ago.

Why, for example, do some facts counts as 'evidence' — and yet we cheerfully dismiss others as atypical, as special cases or just irrelevant? John Stuart Mill asked exactly that years ago (in *A System of Logic*, 1843) saying: 'Why is a single instance, in some cases, sufficient for a complete induction, while in others myriads of concurring instances, without a single exception known or presumed, go such a little way towards establishing an universal proposition?'

More recently, the lurking uncertainty at the heart of scientific method was illustrated by the American philosopher, Nelson Goodman, with his imaginary colour 'grue' — that is green up to a certain time, and blue afterwards. His point was that no tests (prior to that particular time) would distinguish between something that was green, and something that was 'grue'. There's no easy answer to that one, so it is no wonder that we find scientists and experts ignoring the uncertainties instead.

Contemporary debates about the origins of life and the mechanism of natural selection, or more recently about climate change, illustrate how interwoven science is with social values — and how scientists do not *really* proceed from the evidence to formulate a theory, but rather seek the evidence to reinforce a prejudice. Prejudices? But as Thomas Kuhn says in his monumental but ambiguous work *The Structure of Scientific Revolutions* (1962), once scientific theories have become established, they are not only vigorously defended, but rival ideas are fought and suppressed. 'Novelty emerges only with difficulty, manifested by resistance.'

And that is why this is also a book about group-think and how it determines our lives. The writers and philosophy editors Julian Baginni and Jeremy Stangroom once wrote a book called *Do you think what you think you think?* and the answer to that, as they say, is surely no. We think what other people have told us to think, a social reality that stretches from the implicit assumptions of our language net to the explicit lessons of schools and universities. Increasingly, too, we think collectively — as English becomes a shared world language, as the internet replaces both traditional news sources and ages old forms of social interaction, the scope for individuals to really think what they want to think diminishes. Literature, politics, and scientific debates alike are reduced to a kind of hard core of facts, before which we must yield. The range of views and perceptions shrinks.

In my line of work, I mean as a philosophy editor, I often see new books on philosophy and social science, and scarcely a few months go by without a new one appearing warning — as though it was the most original thought in the universe — against irrationality. But this is an old, old story.

The Demon-Haunted World: Science as a Candle in the Dark by the respected astronomer, Carl Sagan, and Ann Druyan (Ballantine Books, 1997) is a good example, still popular despite its age and indeed despite Carl Sagan himself having died just before publication. It shows how strong the desire of readers is to be reassured about what is real science and why all those strange things — Demons, UFOs, the Loch Ness Monster, Big Foot, fairies, and the like — are all foolish nonsense — or pseudoscience, as Sagan calls it. As the book states, 'the siren song of unreason is not just a cultural wrong but a dangerous plunge into darkness that threatens our most basic freedoms.' At least Sagan also urges readers to critically scrutinize information professed by supposed experts operating on the margins, but the book offers no insight into the problem of being taken in by the much more influential, main-stream scientific ones.

A book by the editor of the *Skeptic* magazine, Michael Shermer (which he acknowledges to be following in the footsteps of Sagan's) called *Why People Believe Weird Things: Pseudoscience, Superstition, and Other Confusions of Our Time* (Holt Paperbacks, 2002) attempts to broaden the debate into areas of sociology and human values — as the present work does. Shermer looks, for example, at economic theory in the shape of Ayn Rand's Objectivism, at Holocaust denial, and at the perennial conflict between evolutionary theory and religion.

In fact, books promoting the comfortable certainties of logic and science are everywhere. Francis Wheen huffs and puffs against 'Mumbo Jumbo' and modern delusions; Steven Law gets hot under the collar about vulnerable groups who believe in 'bullshit'; and Sam Harris offers that it is scientists who should henceforth determine 'the moral landscape'. The authors insist that their books are needed because people are still trying to think in ways that are not *right*, not rational. People still believe in astrology, in quack medicine, in anti-scientific theories of all sorts. These books, usually written by journal-ists, are like the barking dogs herding us sheep back into our intel-lectual pens. Yet on the other side, the academic side, there is a sur-prising silence in defence of diversity of opinions, openness to new ideas. It has been a long time since Paul Feyerabend offered his essays on 'methodological anarchism' which disputed so many of the ortho-doxies of science and contemporary society. Likewise it has been a long time since students marched against wars — or even that television networks experimented with truly alternative comedy! In 2015, Paris's *Charlie Hebdo* magazine, which challenged all kinds of political ortho-doxies, met the new intolerance in the most horrible way, with twelve of its journalists shot dead as they discussed cartoons for the next issue. Reporting the event, the *New York Times* declined to reproduce the cartoons, saying that they were potentially offensive! We have become a world-society in which convention has become a vice in which our minds are gripped so tightly that independence of thought and opinion is not only difficult — but dangerous.

Yet what I want to argue here is that, in almost every area you look at, it is excessive orthodoxy, not excessive debate, that is the problem and the danger. From modern medicine — with its mass prescription of mind control drugs—to the grand designs of 'climate control' scientists, working on how to bury carbon dioxide in the ground; everywhere you look—if you can step back a moment from the prevailing orthodoxy—there are good reasons both to fear and to be angry about how the authorities, the experts, the opportunists, and (most of all) the unreflective have not only taken control of our lives (and our futures) but also taken control of the terms of debate. The madness is not, as Charles Mackay supposed in *The Madness of Crowds*, his classic nineteenth-century work on group-think, in the minds of the ordinary people who make up 'the crowds', but in the minds of the elites. It is these who have created hierarchies and structures that can only allow one opinion (invariably a self-serving one) and which stifle the extraordinary power and ability of public debate to find wisdom.

In the chapters that follow, I hope to convince you that there are real debates, real issues that deserve consideration, and that the truth, as Socrates also pointed out so long ago, is that wisdom lies not in learning the thousand and one facts, but rather in realizing how little it is that anyone does, in fact, really know. The irrationality belongs to the ideologues and dogmatists who will allow no dissenting voices.

Why don't giraffes climb trees (if their necks evolved to reach the leaves at the top of the canopy)? Which of these are the most dangerous—salt, butter, peanuts? How long have we got before the Earth overheats? There are lots of possible answers. But the more important thing is to start asking questions.

'How To Defend Society Against Science'—Paul Feyerabend

'Science is just one of the many ideologies that propel society and it should be treated as such... there must be a formal separation between state and science just as there is now a formal separation between state and church. Science may influence society but only to the extent to which any political or other pressure group is permitted to influence society. Scientists may be consulted on important projects but the final judgement must be left to the democratically elected consulting bodies. These bodies will consist mainly of laymen. Will the laymen be able to come to a correct judgement? Most certainly, for the competence, the complications and the successes of science are vastly exaggerated. One of the most exhilarating experiences is to see how a lawyer, who is a layman, can find holes in the testimony, the technical testimony, of the most advanced expert and thus prepare the jury for its verdict. Science is not a closed book that is understood only after years of training. It is an intellectual discipline that can be examined and criticised by anyone who is interested and that looks difficult and profound only because of a systematic campaign of obfuscation carried out by many scientists.'

How to use this book

This is not a book for the obedient, passive reader! It is instead a book for the critical thinker, the curious reader, not to say the contrarian one too.

Throughout, the book encourages the reader to be an active participant—to explore and investigate ideas for themselves. An underlying principle here is that even complex issues are at root rather simple, and not only can be tackled but enjoyed by anyone with an open mind and a preparedness to put aside assumptions.

My aim in writing it is to make a case for scepticism by taking an objective look at broad topics and shining a spotlight on their contentious and sordid pasts, revealing how major flaws and holes were routinely glossed over in the perennial quest for consensus. Controversial themes are tackled, but with sufficient distance and in a spirit of curiosity that it should enable all readers, even those with different personal standpoints, to explore and enjoy them.

Here and there, the book points a finger at today's over-dependence on 'expert' opinion, sharing stories and anecdotes that reveal how expert opinion lurches from one orthodoxy to another as times, circumstances, and fashions change. While some of the chapters included are political with scientific aspects, most are scientific with political aspects. But then one of the arguments here is that the two cannot, actually, be separated, making it only that much more imperative (and difficult) to remain sceptical.

Sceptical issues in everyday 'Life' include debates in medicine and biology, in art, and in buying soap-powder. Grand debates concerning 'The Universe' come from astronomy, physics, and cosmology. The book does not, to be sure, quite look at 'everything' but rather at broad issues in 'theory of knowledge', at the philosophy of science, and at many hotly debated recent debates in society, yet always with a focus on their scientific pretensions to being resolved.

Why don't giraffes climb trees (if their necks evolved to reach the leaves at the top of the canopy)? Which foods are dangerous? How long have we got before the Earth overheats? There are lots of possible answers. But the more important thing is to start asking questions.

Footnotes. The book will contain no technical footnotes, as part of the strategy of being already in plain language, but will include a chapter-by-chapter summary of key sources and suggestions for further reading at the end of the book.

Part I

Chapter One

Tales of Mice and Men

Competing narratives of the origins of life that illustrate the ever-problematic relation of theory and evidence

Subject: Origins

What we're supposed to think:

'Darwinian evolution proceeds merrily once life has originated. But how does life get started? The origin of life was the chemical event, or series of events, whereby the vital conditions for natural selection first came about. The major ingredient was hereditary, either DNA or (more probably) something that copies like DNA but less accurately, perhaps the related molecule RNA. Once the vital ingredient — some kind of genetic molecule — is in place, true Darwinian selection can follow, and complex life emerges as the eventual consequence.'
— Richard Dawkins, *The God Delusion* (2006)

The sensible sceptic's view:

'More than thirty years of experimentation on the origin of life in the fields of chemical and molecular evolution have led to a better perception of the immensity of the problem of the origin of life on Earth rather than to its solution. At present, all discussions on principles, theories and experiments in the field either end in stalemate or in a confession of ignorance.'
— Klaus Dose, biochemist, writing in *Interdisciplinary Science Review, 13* (1998)

To begin at the beginning... well, how did life begin?
It's a fundamental human concern

For indeed, even the simplest of living organisms — like viruses — possess a very complex structure. It has been likened to a tornado rearranging the scrap in a junkyard into a modern jet plane. Yet, one thing can be said; living organisms — including ourselves! — do exist — so how did they come about? A popular explanation in the twentieth century was that the Earth might have been seeded by organic life produced on nearby stars. Tiny bacterial spores, no larger than 0.0002 of a millimetre, blown by no more than the pressure of light waves, might have sent these tiny organisms across space to seed our planet. Implausible though the theory is, it also left the question of just how life started completely unanswered — the problem has simply been displaced to other solar systems and given rather more billions of years to arise. But even to do that requires creative solutions to explain how the spores cope with the destructive effect of radiation in deep space.

Amongst orthodox biochemists today, the assumption is that the mystery of the origin of life is best explained by saying that actually life is not that complicated, but merely the impressive end of a long but otherwise simple process. The celebrated Russian biochemist Aleksander Ivanovitch Oparin put it this way in his classic account of the problem, written in 1936:

> '...the simplest living organisms originated gradually by a long evolutionary process of organic substance and... represent merely definite mileposts along the general historic road of evolution of matter.'

But this approach to the question does not so much solve the mystery as merely relegate the implausible event to a dim and distant past...

Not surprising then that every civilization has its own story or founding myth explaining the very origin of life itself. Today, the story is that liquid water is the key condition for life as we know it, after which, the rapid development of life is the inevitable 'chemical event, or series of events', guided by natural selection. It is merely a matter 'of interest' that during the time that life appeared on Earth the evidence points at similar conditions, including liquid water, on Mars, yet life has, if it exists at all there, not exactly flourished... Anyway, maybe there is life on Mars... underground.

Luigi Galvani (1737–1798) was an Italian physician, physicist, anatomist, and philosopher who studied medicine and practised as a doctor in Bologna. Galvani became known for his experiments revealing the secrets of life, the most notable example taking place on 6 November 1787 when Galvani discovered that electrical current could cause movements in the limbs of dead animals and nerve tissue. He demonstrated this through experiments with freshly killed frogs in which an electric charge was enough to make their legs twitch. In fact, his explanations for the phenomena were wrong, but at the time most scientists were excited by the idea, and his work stimulated many new lines of investigation, in physiology and in electricity – and indeed in literature. Indeed, it is thought that Mary Shelley's classic work, *Frankenstein*, in which the monster is brought to life by an electrical charge, owes its starting point to Galvani. Within the scientific community, though, some ridiculed Galvani's work, which was unashamedly rooted in the goal to *create* life and in so doing to upturn the natural order (the strategy set out in Mary Shelley's *Frankenstein*). The works of the Italian experimenter also illustrate the important fact that science is rarely content to be a rational endeavour but rather is rooted in an attempt to rationalize life.

And that is an ancient story. For a rather different example, recall the that indigenous Australians speak of the 'dreamtime', in which supernatural beings arose from the ground, taking on the shapes of animals, such as Wallaby Dreaming and Emu Dreaming. In their songs and chants, they record how these strange beings created the landscapes, the mountains, and rivers – usually through battles and conflicts. In this creation myth, the spirits and characters of the animals are the building blocks that led to the Earth as we see it today... It's a beautiful and powerful way to understand the world...

But today people prefer the scientist's kind of tale, which in this case revolves around the spontaneous generation of plants and animals out of a kind of chemical soup.

It is not a new story: in ancient China it was thought that aphids were spontaneously generated from bamboos. In India, sacred documents mention the spontaneous formation of flies from dirt and sweat. And in Babylonia, inscriptions explain that mud from canals was able to generate worms.

Yet, as science has become more sophisticated, it has firmly demolished more and more cases of the 'spontaneous generation' of life. At the same time, however, despite their very best efforts, to date no scientist has managed to create life out of just chemical ingredients. And that's an ideological hurdle for the advocates of evolutionary theory.

Nonetheless, for most scientists today, the story is the same as it was for most of the ancient Greek philosophers: life arose naturally out of the elements of the universe, is eternal, and will appear spontaneously whenever and wherever the conditions are favourable.

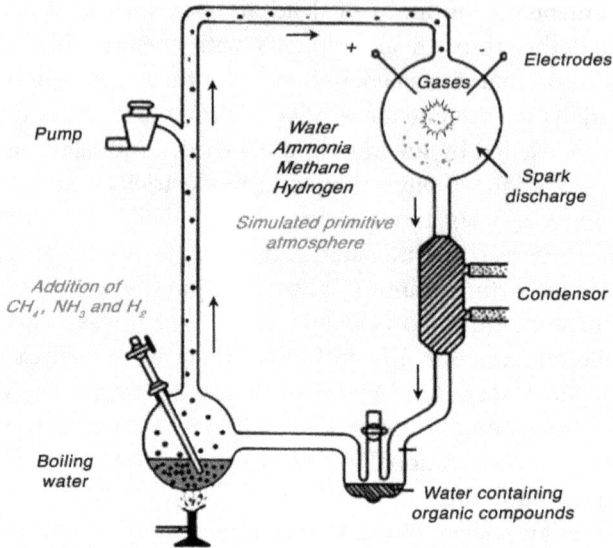

Figure 1. Stanley Miller's 1953 laboratory model of how life begun, for which he was nominated for (but did not receive) a Nobel prize.

***

Thales seems to have started the fashion for explanations of the natural world that are also rooted in… the natural world. For Thales, everything was ultimately made out of just one thing—and this was water. As Bertrand Russell says, the statement that everything is made out of water is by no means a foolish one. Wrong, yes, but foolish, no. Russell and many other natural philosophers present Thales as offering a radical chemical deconstruction of the superficial complexity of the world around us, even though, in all likelihood, Thales was probably using water as a metaphor—referring to the process of change that Eastern thinkers were already promoting and exploring.

The same seems not to be true, however, of Anaximander, Thales's successor, who elaborated on the theory in the fifth century BCE, writing (and this is said to have been one of the first books) that living creatures were first formed in the 'wet' as a result of the heat of the Sun, and that these first creatures were unlike anything we have now.[1] Anaximander's own pupil, Anaximenes, then fine-tuned the approach

---

1 Anaximander, however, disagreed with Thales that everything was ultimately water, arguing that such an arrangement would undermine the 'balance' of the universe. Instead, he seems to have thought that everything must be a mixture of the primary elements—fire, earth, air, and water—which in turn were all made out of some mysterious other element.

by suggesting that air was the crucial element that imparted life, not to mention motion and thought, and that it was the Sun's rays acting on the primordial slime, a mixture of earth and water, in a kind of evaporation, that directly formed plants, animals, and human beings.

All the great thinkers of the time, such as Xenophanes, Parmenides, and Anaxagoras, agreed with the general approach, although this last speculated that the seeds of plants existed in the air from the beginning, while those of animals and humans came from eggs in the heavenly ether. Empedocles even suggested (in another early kind of evolutionary theory) that there might also have been various trials of combinations of parts of animals that spontaneously arose, but that only the successful forms managed to reproduce. Nonetheless, it seemed that all the experts—Thales, Democritus, Epicurus, Lucretius, and, of course, Plato—were thinking along the same lines.

But it was Aristotle who dressed the different claims up as a truly *scientific* theory. It is Aristotle who says that plants spontaneously appear out of the earth, fish appear out of water, and that animals are a consequence of a power in the air. As for fire, why it generates spirit beings, such as inhabit the moon. This fine theory then traversed the Middle Ages to impress Renaissance thinkers like Newton, Descartes, and Francis Bacon.

It is in the *History of Animals* that Aristotle details how life begins:

> '...some spring from parent animals according to their kind, whilst others grow spontaneously and not from kindred stock; and of these instances of spontaneous generation some come from putrefying earth or vegetable matter, as is the case with a number of insects, while others are spontaneously generated in the inside of animals out of the secretions of their several organs.' (539a18–26)

Ever the keen biologist, Aristotle observed directly crabs generating in damp sand, frogs developing in slime, and even mice in wet soil. He offers particularly detailed reports on insects, many of which, he says:

> '...are not derived from living parentage, but are generated spontaneously: some out of dew falling on leaves, ordinarily in spring-time, but not seldom in winter when there has been a stretch of fair weather and southerly winds; others grow in decaying mud or dung; others in timber, green or dry; some in the hair of animals; some in the flesh of animals...' (551a1–10)

Drawing on his observations of marine life, he affirms:

> 'it is quite proved that certain fishes come spontaneously into existence, not being derived from eggs or from copulation. Such fish as are neither oviparous nor viviparous arise from one of two sources: from mud, or from sand and from decayed matter that rises thence as a scum; for

instance, the so-called froth of the small fry comes out of sandy ground.'
(569a21–569b3)

And in a text known specifically as the *Generation of Animals,* he summarizes:

> 'Animals and plants come into being in earth and in liquid because there
> is water in earth, and air in water, and in all air is vital heat so that in a
> sense all things are full of the animating force. Therefore living things
> form quickly whenever this air and vital heat are enclosed in anything.
> When they are so enclosed, the corporeal liquids being heated, there
> arises as it were a frothy bubble.' (Book 3, ch. 11)

Such was Aristotle's prestige and influence that for the next two thousand years the only challenge to the theory was over whether it might be *only* simple bodies and organisms that could spontaneously generate, or whether it was also more complex animals – and even humans. One such fierce debate that preoccupied scholars for centuries concerned the *Legend of the Lamb-Plant.* This mysterious organism sprang from a seed 'like that of a melon, but rounder', and produced a fruit or seed-capsule which eventually burst open when ripe to reveal – a little lamb within it! The wool of this little lamb was described as being 'very white'.

One nineteenth-century chronicler recalls how, if tended carefully, the little lamb 'would grew to a height of two and a half feet and had a head, eyes, ears, and all the parts of the body of a newly born lamb. It was rooted by the navel in the middle of the belly, and devoured the surrounding herbage and grass'.

This peculiar story captured the imaginations of not merely humble people (aware of the Christian significance of little white lambs), but the most celebrated writers, philosophers, and otherwise very respectable scientific men of the time.

No wonder that Darwin's grandfather, Dr Erasmus Darwin, writing at the beginning of the nineteenth century, was still prepared to write in his scientific poem *The Temple of Nature* (1802) (much appreciated at the time but only rarely later, as one wag put it), that:

> 'Hence without parents, by spontaneous birth
> Rise the first specks of animated earth.'

Erasmus D. had probably heard that the general theory had been experimentally confirmed (in the middle of the seventeenth century) by such as Johannes Baptista van Helmont. Johannes, a Flemish physician and alchemist, had successfully spontaneously generated mice from wheat grains and a sweat-stained shirt incubated in a warm dark closet. Being a serious and careful type, he noted that the mice seemed in *every respect* identical to more conventional ones (obtained by procreation), such as those running around his house.

Nonetheless, as with most successful theories, critics had also appeared on the scene. In England, William Harvey nibbled away at the claims in his *De Generatione* (1651), in which he said that many cases of apparently spontaneous generation were due to invisible seeds being scattered and dispersed through the air, and coined a later much quoted phrase *ex ova omnia* (everything comes from eggs) instead.

Indeed, in 1668, Francisco Redi, a Tuscan physician, cast doubt on the spontaneous generation of 'worms' (what we call 'maggots') from rotting meat. He carried out several rather unappetizing investigations into the matter, following the development of larvae from eggs, on different meats including lion meat, lamb, fishes, and snakes. He found that while worms reliably appeared on meat left in the air, when the meat was covered with a thin layer of muslin, none did. This showed, he thought, that the worms did not, contrary to Aristotle, arise spontaneously, but probably came from eggs laid by adult flies. In *Experiments on the Generation of Insects* he wrote:

> 'I began to believe that all worms found in meat derived from flies and not from putrefaction. I was confirmed by observing that, before the meat became wormy, there hovered over it flies of that very kind that later bred in it. Belief unconfirmed by experiment is vain. Therefore I put a snake, some fish, and a slice of veal in four large, wide-mouthed flasks. These I closed and sealed. Then I filled the same number of flasks in the same way leaving them open. Flies were seen constantly entering and leaving the open flasks. The meat and fish in them became wormy. In the closed flasks were no worms, though the contents were now putrid and stinking. Outside, on the covers of the closed flasks a few maggots eagerly sought some crevice of entry.'

However, renewed support for the Ancients came from philosophy in the form of Gottfried Willhelm Leibniz (1646–1716). The celebrated metaphysician, mathematician, and scientist asserted that there were fundamental 'living molecules', called *monads*, and it was from these that all things sprang. However, he did think that there was a scale of complexity, and that only very simple organisms were comprised directly of the living *monads*.

As so often, attitudes were shifted by new technologies. So it was in this case that, just six years after Fransisco Redi's experiments with flies' eggs, a Dutch scientist, Anton Van Leeuwenhoek, observed microorganisms for the first time through the new-fangled contraption known as the microscope (this one of his own making). Soon, microorganisms were being found everywhere. Even so, it wouldn't be until about 200 years later when Louis Pasteur (responding to a competition) designed a rigourous experimental set-up, to keep micro-organisms out of laboratory flasks, that the idea that life was continually emerging from everyday chemical processes collapsed.

It was in 1862 that the French Academy of Science offered a prize of 2,500 francs (a considerable fortune in those days) to whoever could shed 'new light on the question of so-called spontaneous generation'. Pasteur would eventually win it for his essay *'Mémoire sur les corpuscules organisés qui existent dans l'atmosphère'*, which was then published in the Academy's *Annales*. In this essay he describes a series of elegant experiments designed to show how dust carrying germs could explain cases of 'life' otherwise apparently arising spontaneously in sealed flasks of sterilized broth.

As part of his practical demonstrations, Pasteur had flasks made with a series of differing shapes designed to allow the movement of air but not the entry of dust. The liquid inside remained clear for months. As one biographer noted:

> 'The observer had a choice between only two hypotheses: placing the origin of germs either in solid particles (fragments of wool or cotton, starches) that float in the atmosphere, or in spores of moulds or the eggs of *infusoria*.'

As to which, Pasteur responded succinctly: 'I prefer to think that life comes from life.' Louis Pasteur is rightly one of the great names in the history of science and medicine and this is in large part because of his conclusive demolition of the, until then, dominant concept of spontaneous generation and the establishment in its place of the germ theory of disease. A physicist as well as a chemist, he was the first to explain the organic basis and control of fermentation, as well as to develop the processes of *pasteurization* and sterilization. It was as he followed his research increasingly into bacteriology that he not only discovered a number of disease-producing organisms but developed vaccines to combat them. We owe to him both our understandings and our practical responses to a myriad of nasty diseases: rabies, diphtheria, anthrax —to mention just a few. In this way, Pasteur single-handedly saved countless human lives.

Yet, it is instructive to note, in his lifetime Pasteur was a hate figure and the object of intense opposition by almost the entire scientific establishment, precisely because of his opposition to the Ancients and their theories of 'spontaneous generation', as well as, to a lesser extent, because of his resistance to the new ideas of his English contemporary, Charles Darwin. Only in the face of repeated experimental and analytical demonstrations did biologists and medical scientists begin, reluctantly, to give up their cherished ideas.

Another triumph, if rather belated, of scientific method over philo-sophical speculation? Not quite. The all important 'victory' in the Academy's 1864 competition was a long and error-strewn process which hinged on the rival sampling of air (from a balloon by Pasteur, from the ceilings of cathedrals by his rival, Felix Pouchet, and from high mountaintops by both). It was all overseen by a blatantly biased committee of the Academy and even then was only settled in Pasteur's favour after Pouchet walked off in protest, complaining of bias and misprocedure. But then, because the theory of spontaneous generation was considered to diminish God's role in creation,[2] it was essentially a very 'political matter'. After his victory, Pasteur, a devoutly religious man himself, became a hero in Catholic French society.

Strictly speaking, Pouchet (who was religious himself) *should* have triumphed as he had proved that broth made with hay could generate life even when boiled. Actually, as became known later, this is only because the spores on hay are heat resistant. Nonetheless, had he stayed in the competition, he could very well have won 'on the evi-dence'. Indeed, when Pasteur's notebooks were made available (only in the 1970s), it emerged that the great man repeatedly ignored unwanted, positive results in experiments, claiming that they were due to error rather than spontaneous generation. In fact, only a minority of his experiments gave the results he anticipated and desired.

So what did Pasteur prove? Did he prove that life can never come from non-living things? Not at all, as this sort of negative can never be proved experimentally, only theoretically—philosophically, one might say… What he showed, rather, was that it was *highly unlikely* that modern living organisms arose from non-living organic material.

Pasteur's contribution was instead, as Darwin put it in an essay, to show that a 'mass of mud with matter decaying and undergoing com-plex chemical changes is a fine hiding-place for obscurity of ideas'. And after Pasteur's demolition of 'the old doctrine of spontaneous genera-tion', it was easy for Darwin to convince others that even if it were true historically that 'there must have been a time when inorganic elements alone existed on our planet', it was equally true that 'our ignorance is as profound on the origin of life as [it is] on the origin of force or matter', and hence to deny that the theory of evolution needed to pro-vide an explanation for it, far less to claim that life continuously arises.

---

2   For the other side, the celebrated pathologist and cytologist Rudolph Virchow (1821–1902) considered spontaneous generation as 'heresy, or devil's work' in 1855, and asserted that: 'Life will always remain something apart, even if we should find out that it is mechanically aroused and propa-gated down to the minutest detail.'

Chapter Two

Discarding
Fossilized Theories

So how do you spot a fossilized theory?
Further examining the political nature of science

Subject: Evolutionary Theory

What we're supposed to think:

'The principles of evolution have been tested repeatedly and found to be valid according to scientific criteria. Evolution should be part of the pre-college curriculum; it is the best scientific explanation of human and nonhuman biology and the key to understanding the origin and development of life.'
— *Statement on Evolution and Creationism*, adopted by the American Anthropological Association, Executive Board, April 2000

What you're not to say:

'A fair result can be obtained only by fully stating and balancing the facts and arguments on both sides of each question; and this cannot possibly be here done.'
— Charles Darwin, *The Origin of Species* (Introduction), commenting on the question of the truth or otherwise of his theory

You don't need to believe in God to disbelieve in evolution. Indeed, its proponents, contrary to what one might think appropriate, seem to rely more on blind faith than rational arguments in support of the theory — because there are significant holes in the edifice of Darwinism, and plenty of dodgy arguments needed to support it.

Darwin himself was perfectly prepared to accept doubts about his theory. New editions of his book included many modifications which, by the time the last edition was published, had become frequently contradictory. Yet even as Darwin accepted the complexity of the issue, the public had already become convinced that natural selection was an idea that solved everything.

To judge by the newspapers today, the debate over the theory of evolution — or natural selection to be more precise — is conducted between Bible-wielding bigots on the one hand, fixated on the start of the universe 3001 years ago, and knowledgeable but weary scientists, equipped with sophisticated carbon-dating technologies tracing life back billions of years. Yet the truth is never so simple and there is an important debate to be conducted somewhere on the middle ground that often gets lost during all the fiery denouncements. And it's a debate in which both sides have made some uncomfortable alliances.

In fact, scientific opposition to Darwin's theory is still there, clinging to existence in rocky crevices like the French Academy of Sciences.[1] There are a few fierce, evolutionary-theory-eating critics within the history of biology, such as Ernst Chain (1906–1979), considered to be one of the founders of the field of antibiotics (for identifying the structure of penicillin, and isolating its active element). When it came to Darwin's theory of natural selection, Chain had harsh words. He found it to be 'a very feeble attempt' to explain the origin of species, based on assumptions so flimsy that 'it can hardly be called a theory'. The central notion of chance mutations resulting in complicated adaptations he

---

[1] There are a few critics. The French biologist Pierre P. Grasse, sometime President of the French Academy of Sciences, flatly called Darwinism 'a pseudo-science' that is 'either in conflict with reality or cannot solve the basic problems' and, in recent years, the British-Australian molecular biologist Michael Denton has weighed in with a potboiler called *Evolution: A Theory in Crisis* (1986), in which he seeks to demonstrate that recent developments in molecular biology cannot be reconciled with Darwin's ideas. A similar claim that certain features of biochemistry, such as the mechanism of blood clotting, are too complex to have arisen 'by chance' is made by the American academic Michael Behe in his book, this time called *Darwin's Black Box*. Behe says he used to accept Darwinism, and now has no idea what could explain the 'irreducible complexity' of the natural world. But to his many critics, he is a creationist. Wicked man!

dismisses as an 'hypothesis based on no evidence and irreconcilable with the facts'. Noting, however, that most of his colleagues happily accepted the theory, he said he was amazed that such a gross over-simplification of an immensely complex and intricate mass of facts could be 'swallowed so uncritically and readily, and for such a long time, by so many scientists without a murmur of protest'.

Yet so, without a murmur, it has. Today, most scientists and philo-sophers still call themselves 'evolutionists' since they recognize that all life forms share basic genetic material and so may be descended from a single ancestor; even if they are frank about not being able to explain how this happened. Some speculate (uselessly) that evolution was somehow 'pre-programmed' in the DNA.

But how can we discount the tenets of our beloved theory of evolu-tion? It is the best explanation for the world around us we've come up with, after all. Or so the thinking goes. And all thanks to Darwin's careful, scientific studies...

Actually, the idea that life forms have changed over long periods of time has been around a lot longer than Darwin, certainly at least since the ancient Greeks. Noting that the fossils of shells can be seen in rocks, the Ancients surmised that the world had once been covered in seas, and hence that human beings were descended from fish that had later taken to dry land. (How many times have we heard that one?) It is Anaximander, who lived in the sixth century BCE — that is some centuries before Socrates and Plato — who should probably be credited with being the first person to explain the development of life on Earth as a kind of evolutionary process.

If he ever defined the process precisely, the detail has been long lost, but it has been generally taken that the primeval state was a time of swirling chaos as all the opposites — hot and cold, wet and dry — battled with each other. In the process all the shapes and differences which are found not only on Earth but on 'all the worlds', of which Anaximander believed there must be very many, were created. (Curiously, recent so-called anthropomorphic arguments in support of Darwin's approach also rely on a near infinite number of Earth-type worlds to explain the otherwise implausible 'coincidence' of life.)

And on this Earth, life began in the seas. Anaximander supposed that life, in the form of fish, must have spontaneously arisen in the moisture that originally covered the whole of Earth.[2] Man himself came into being only after a long series of transmutations as other kinds of animals. For this, even though Anaximander only hints at a theory of

---

2   Very much in the manner of today's biochemists, albeit these typically settle for the spontaneous forming of various elementary amino acids.

natural selection, some people consider him to be evolution's most ancient proponent. Indeed, the theory of an aquatic descent of man was re-conceived centuries later and graced with the name 'the aquatic ape hypothesis'.

One surviving fragment of Anaximander's writings offers:

'Whence things have their origin,
Thence also their destruction happens,
According to necessity.'

In just 13 words, that's Darwin's theory in a nutshell! Not to forget, a good century before Darwin published his *Origin of Species* (1859), the French zoologist Buffon also proposed a mini-theory of evolution. That's why a number of thinkers in France and England, including Darwin's paternal grandfather, Erasmus, had begun to champion the idea. At the time that Darwin wrote his book, the notion that man was descended from some early fish-like creature was very much the talk of London dinner tables.[3] That life had evolved was not in doubt. It was hardly Darwin's 'discovery'. The question was how.

What Darwin brought to the feast was his theory of natural selection. Here was a splendid new idea that was so simple that Thomas Huxley, the great Victorian biologist, exclaimed (of himself), 'How stupid not to have thought of that!'

Conventionally, Darwin is supposed to have arrived at his insight as a result of observing animals on the Galapagos Islands during the voyage of the Beagle. But the real root seems to have been a famous essay, written in 1798 by the English economist Thomas Malthus, 'On Population'. It is here that Malthus, obsessed with the dangers of over-population, warns that humanity is locked in a 'struggle for existence' in which only the fittest survive, and that:

'…it follows that any being, if it vary ever so slightly in a manner profitable to itself… will have a better chance of survival, and thus be naturally selected.'

It is known that Darwin read Malthus's essay, was much impressed by this notion, and merely expanded this theory of 'struggle' among humans to the wider sphere of the plants and animals. However,

---

3   Here's how Darwin put it in *Variorum*, 702, 306.10: 'It is probable, from what we know of the embryos of mammals, birds, fishes, and reptiles that [these animals] are the modified descendants of some one ancient progenitor, which was furnished in its adult state with branchiae, had a swim-bladder, fours simple limbs, and a long tail fitted for an aquatic life' (quoted in 'Embryology and Morphology', an essay by Lynn K. Nyhart in *The Cambridge Companion to the Origin of Species*, edited by Michael Ruse and Robert Richards, Cambridge 2009, p. 214).

Darwin did add another element too, which was that the offspring of most species vary slightly from their parents—some can run faster, some have sharper teeth, and so on—and that thus nature will inevitably favour some adaptations over others in the 'struggle for survival'. This new Darwinian evolution, then, is a two-step process: random mutation is the raw material, natural selection is the guiding force.

Darwin spent much time observing animal breeders at work in and around London. He recorded how selective breeding of pigeons could result in varieties with different features—wider wingspans, longer beaks, and so on. It seemed to him entirely logical to suppose that since the particular characteristics of species could be changed by breeders relatively easily, nature herself could over thousands of millions of years do far more. Indeed, nature could easily turn fish into men.

But then, Darwin had as yet no examples of one species that had changed into another. His argument was entirely theoretical; philosophical indeed, based on thought experiments, not empirical observation. He assumed small changes could add up, over time, into big changes—and hoped to find the evidence later.

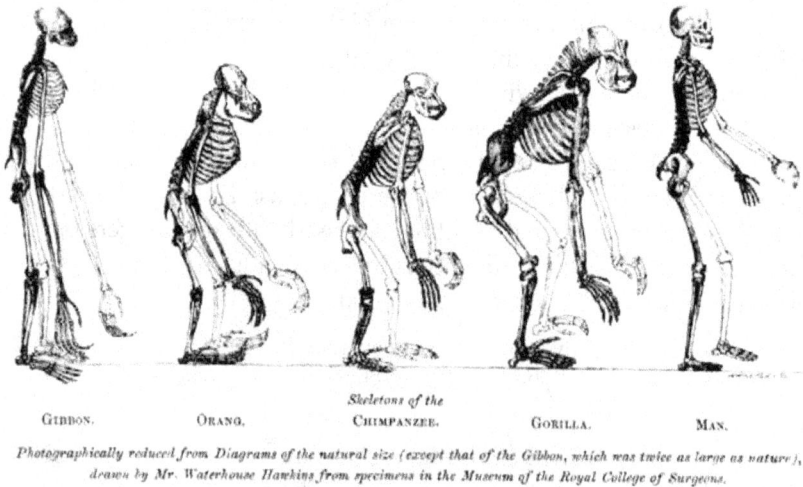

Skeletons of the
GIBBON.			ORANG.			CHIMPANZEE.			GORILLA.			MAN.

Photographically reduced from Diagrams of the natural size (except that of the Gibbon, which was twice as large as nature),
drawn by Mr. Waterhouse Hawkins from specimens in the Museum of the Royal College of Surgeons.

Figure 2. The march of Man.

First of all, he consulted the fossil evidence. The theory seemed to require innumerable transitional forms linking past and present species. Darwin spent many happy hours hacking at rock faces, but to no avail. Even today, much digging later, the rocks show only that species seem to appear abruptly in a fully developed state and change little or not at all before disappearing.

And it seems that the higher you go in the taxonomic order, the bigger the gaps in the record. All 32 orders of animals, for example, seem to appear effectively out of nowhere. Most biologists no longer attribute these systematic gaps to an 'imperfect' record. They accept that the 'gaps' are real and surmise that they reveal something significant about how life 'actually' developed on Earth: and that, at least, was not at all as Darwin imagined.

The blank spots on evolutionary 'tree' charts occur at just the points where, according to Darwin's theory, the crucial changes had to take place. It offends the critics that the direct ancestors of all the major orders – primates, carnivores, and so forth – are completely missing. There is no fossil evidence for a 'grandparent' of the monkey, for example. 'Modern gorillas, orangutans, and chimpanzees spring out of nowhere', writes palaeontologist Donald Johansen. 'They are here today; they have no yesterday.'

High hopes for the linking of two species came with the dramatic discovery of the Piltdown Man fossils back in 1912. The Piltdown Man was the discovery, not of a whole man to be sure, but of a skull and jawbone combining characteristics of modern man with those of the great apes. The find thus appeared to provide one 'missing link' in Man's evolutionary record. The experts, in this case 'all the world's archaeologists', accepted the bones as being of a hitherto unknown form of early human. It appears no one bothered to examine them closely, assuming that other scientists had thoroughly investigated and vetted it. As a result, it was not until 1953 that the hoax was uncovered. It turned out that the skull of a 'modern' human had simply been buried along with the jawbone of an orangutan.

Trying to Understand Worm Brains

In their search for intermediate stages in the evolution of *Homo sapiens* scientists now think the 'missing link', or more accurately one of the 'missing cousins', is none other than the marine ragworm, a humble creature with ancient roots that crawled around the bottom of the sea hundreds of millions of years ago—and still does.

Recent research has found that the ragworm has something resembling the cerebral cortex, the part of the brain that in humans is responsible for our highest kinds of thinking—creative, analytical, and, yes, philosophical too.

Commenting on the similarities, exposed after miniscule examination of the chemical character of tiny bits of worms, one researcher, Detlev Arendt of the European Molecular Biology Laboratory, said: 'You can say that the topography is so similar that the human and worm must come from a common ancestor.' Do ragworms need to think creatively—to philosophize? Well, not exactly. The theory is rather that they need to smell out food, and that it is in this biological prompting that our higher mental facilities originate.

Darwin was aware that the fossils of his day lacked many transitional forms. He accordingly entitled one chapter on the subject 'On the Imperfection of the Fossil Record', and voiced the conviction that future digging would fill the gaps and show the gradual evolution of species.

Ever since, whole armies of palaeontologists have searched higher and higher mountains and dug in deeper and deeper holes for evidence of past 'link' species. Alas, the rocks still show exactly what they showed in Darwin's day: species coming and going (mostly 'going', as 99 out of every 100 species are extinct) in a mysterious theatre performance structured in distinct acts.

The curtain rose on the show with a subtly understated performance by bacteria, 3.5 billion years ago, hesitantly followed by cameo appearances by blue-green algae and a few other oddities. (What evolutionists call the 'Last Universal Ancestor' lived then.) However, nothing resembling 'animal life' evolved until all of a sudden (fine dramatic twist) comes the Cambrian explosion, about 550 million years ago. The fossil record reveals a sudden profusion of complex life forms: jelly fish, trilobites, mollusks, for which there are no discernible ancestors in the earlier rocks. Even Richard Dawkins, obedient disciple as he is of Darwin, struggles to find an explanation for this sudden riot of new life: 'It is as though they were just planted there, without any evolutionary history', he writes. 300 million years later, another sudden explosion of species resulted in a rush of flowering plants, an occurrence which puzzled Darwin, and continues to perplex botanists today. As for Giraffes, elephants, wolves, and all species; they all simply burst upon the scene around seven million years ago... Elephants, which evolutionary theory treats as overgrown pigs, are particularly significant in that they use their trunk to manipulate tools, and have complex social behaviours. Speaking of clever animals, *Homo sapiens* himself appears quite suddenly on the scene too, a mere blink of evolutionary time away, but already quite different from all the other species. He draws pictures in caves, and cooks — and talks.

Elephants today still use their trunks to manipulate branches to get at food, and humans still talk and paint, even if they nowadays mostly watch TV. Which underlines another problem for the theory: once a species appears, it seems to stubbornly remain the same. Geneticists have spent years observing the mutations of fruit-flies and have found that no matter how many little changes you add together, watching over thousands and thousands of fruit-fly generations — a fruit-fly is still a fruit-fly. Likewise, bees preserved in amber from forty million years ago are almost identical with living bees.

Fanciful pictures of how giraffes getting longer and longer necks in their pursuit of tasty acacia leaves, or of apes progressively turning into *Homo sapiens,* via club-wielding Neanderthal cavemen, are just that — fanciful. Giraffes, it turns out on closer inspection, spend most of their time, and certainly the critical summer period when food is in short supply, eating low-lying bushes. Their magnificent long necks actually make eating difficult. (That's also an argument against a heavenly 'designer' too incidentally!) Perhaps giraffes would have done better to have evolved claws and start to climb trees. Similarly, nowadays the evidence points not at Neanderthals metamorphosing into *Homo sapiens,* but at the two 'hominids' living alongside, albeit probably not very profitably. (These days, archaeologists speculate that thinking man ate his flower-garland wearing cousin…)

Or take the problem of 'all or nothing' traits like wings. Either the wings work well so you can fly, or they are of little use. Advocates of natural selection, like Richard Dawkins, insist that gradual random mutations resulting in bigger and more powerful 'wing-things' can explain how birds evolved from earth-bound species. Dawkins' insistence on the universal reach of Darwin's theory obliges him to make some strange claims. To the objection that specialized organs like wings (or eyes) seem to be useless until after many supposed cycles of evolutionary refinement, he says that animals might have found 'half a wing' useful even before they could fly. He offers hopefully that half-wings might slow the descent of any unfortunate animal if they fell from a tree. Indeed, he says, 51% of a wing would significantly advantage such an unfortunate falling animal over another of the same species with only 50% of a wing. There is only one question left unanswered, which is how all those flightless birds and other animals, with only two feet and useless wings, got up the tree in the first place!

Darwin's Thought Experiments

At one point in *Origin of Species*, Darwin offers:

'In order to make it clear how, as I believe, natural selection acts, I must beg permission to give one or two imaginary illustrations. Let us take the case of a wolf which preys on various animals, securing some by craft, some by strength, some by fleetness; and let us suppose that the fleetest prey, a deer for instance, had from any change in the country increased in numbers, or that other prey had decreased in numbers, during the season of the year when the wolf is hardest pressed for food.'

Darwin's answer is emphatic:

'I can under such circumstances see no reason to doubt that the swiftest and slimmest wolves would have the best chance of surviving, and so be preserved or selected.'

I suppose we must forgive Darwin his prejudice against fat wolves. But there is another aspect of the example less easily overlooked. Fleeming Jenkins, of Edinburgh University, at once pointed out that there is something dubious with the assumption that such traits could be passed on. Nature tends to 'iron out' individual differences, not to promote them. If the swiftest, slimmest wolf is a rare mutant, then that trait will in fact die out as the inevitable result of interbreeding, however advantageous.

And so, for a moment, it looked like the theory of natural selection, after a relatively short lifespan, was already extinct. But, returning to his example in the later editions of the *Origin of Species*, Charles Darwin makes some small but significant changes, shifting the emphasis to the collective effect rather than the individual one. He now writes:

'Under certain circumstances individual differences in the curvature or length of the proboscis etc, too slight to be appreciated by us might benefit a bee or other insect, so that certain individuals would be able to obtain their food more quickly than others, and the communities in which they belonged would flourish and throw off many swarms inheriting the same peculiarities.'

So, struggle over. A thought experiment led Darwin to significantly change and improve his theory. In fact, it had to evolve in order to survive.

Since the fossil record gives no proof of the gradual transformation of species which Darwin's theory demands, the only other place to look is breeding experiments. This is an idea almost as old as Anaximander's. Plato describes in the *Republic* how dogs can be bred for certain characteristics, and speculates that the same must be true for humans too. But here too the evidence also goes against Darwin. All breeders report the same experience: if they try to go too far in one direction, the animal or plant in question either becomes sterile or reverts back to type.

Instead, today the talk is of 'punctuated equilibrium', according to which small groups of animals mutate rapidly into 'hopeful monsters' who then replace the old herd. Because the changes occur so quickly, we should not be surprised that there is no fossil evidence. Yet even as punctuated equilibrium offers natural selection a prop, it is at the same time a refutation of Darwin, who said that his theory would break down completely 'if it could be demonstrated that any complex organ existed, which could not possibly have been formed by numerous, successive, slight variations'.

That is why many people prefer to cling to old-fashioned Darwinism for the origin of species. For these true-believers the engine of species creation is eeny-weeny DNA copying errors, which add up over millions of years to 'evolution'. These loyalists brush away the gaps in the fossil record and other doubts.

Why, in the face of so much negative evidence, does Darwin's theory maintain its hold over scientists and educators? Perhaps because it is also a powerful general theory, with social and political dimensions. *Origin of Species* has had a profound effect not merely on understandings of biology and nature, but on views of human societies and morality. In Europe, the theory challenged the pretensions of the Church to have the only explanation for how the world came to be the way it is, and subtly shifted power with the elite. For many, it seemed to promise a similarly progressive trend to political systems, for some this was to be manifested through the destructive power of revolution. Nazism and Marxism both were, in the minds of their founders, simply Darwin's theory applied to politics.

Connecting Darwin to unpopular political theories may seem like a dodgy argumentative tactic. Yet Darwin openly extended his theory to cover the human race and therefore challenged many social, ethical, and psychological assumptions. In *Origin of Species*, he argues that moral values are just another form of randomly generated behaviour whose effect was to improve the chances of species preservation.

In a key passage in Chapter 21, the 'General Summary and Conclusion', Darwin writes:

'The moral nature of man has reached its present standard, partly through the advancement of his reasoning powers and consequently of a just public opinion, but especially from his sympathies having been rendered more tender and widely diffused through the effects of habit, example, instruction, and reflection. It is not improbable that after long practice virtuous tendencies may be inherited. With the more civilised races, the conviction of the existence of an all-seeing Deity has had a potent influence on the advance of morality. Ultimately man does not accept the praise or blame of his fellows as his sole guide, though few escape this influence, but his habitual convictions, controlled by reason, afford him the safest rule. His conscience then becomes the supreme judge and monitor. Nevertheless the first foundation or origin of the moral sense lies in the social instincts, including sympathy; and these instincts no doubt were primarily gained, as in the case of the lower animals, through natural selection.'

Reducing morality to self-interest and survival skills was always controversial, but Darwin didn't stop there. If *Origin* is his most admired work, he wrote other less celebrated ones too, such as *The Descent of Man* (1871), which is full of political advice:

'With savages, the weak in body or mind are soon eliminated; and those that survive commonly exhibit a vigorous state of health. We civilised men, on the other hand, do our utmost to check the process of elimination. We build asylums for the imbecile, the maimed and the sick; we institute poor-laws; and our medical men exert their utmost skill to save the life of every one to the last moment. There is reason to believe that vaccination has preserved thousands, who from a weak constitution would formerly have succumbed to small-pox. Thus the weak members of civilised societies propagate their kind. No one who has attended to the breeding of domestic animals will doubt that this must be highly injurious to the race of man. It is surprising how soon a want of care, or care wrongly directed, leads to the degeneration of a domestic race; but excepting in the case of man himself, hardly anyone is so ignorant as to allow his worst animals to breed.' (1st edition, *The Descent of Man,* pp. 168–9)

Scarcely surprising that when Hitler chanced upon such snippets of Darwin's writings he was a great fan. In fact, the, ah... German visionary offers his own variation on the theory in *Mein Kampf.* Here, he explains that: 'A stronger race will drive out the weak, for the vital urge in its ultimate form will, time and again, burst all the absurd fetters of the so-called humanity of individuals, in order to replace it by the humanity of Nature which destroys the weak to give his place to the strong.'

Is that true, let alone fair, to Darwin's idea? Well, yes and no. Darwin promptly distances himself from his own conclusion, rejecting it on grounds of common humanity. Mind you, he excludes 'the weaker and inferior members of society' from this protection.

'The aid which we feel impelled to give to the helpless is mainly an incidental result of the instinct of sympathy, which was originally acquired as part of the social instincts, but subsequently rendered, in the manner previously indicated, more tender and more widely diffused. Nor could we check our sympathy, even at the urging of hard reason, without deterioration in the noblest part of our nature... Hence we must bear without complaining the undoubtedly bad effects of the weak surviving and propagating their kind; but there appears to be at least one check in steady action, namely the weaker and inferior members of society not marrying so freely as the sound; and this check might be indefinitely increased, though this is more to be hoped for than expected, by the weak in body or mind refraining from marriage.'

So where does all that get us? Is Darwinian evolution a great idea — or not? It's probably no good asking those scientists brought up on it, who remain largely in its thrall, like nervous rabbits caught in a car's headlights. Instead, let's just put it this way. *The golden rule of science is that theories are open to revision, and debate.* Darwinian evolution, theories of evolution of all kinds, should be no exception.

Chapter Three

The Brain Doctors

Are psychiatric disorders fact or fiction?
Highlighting the transient nature of certain
kinds of scientific and medical consensus

Subject: Clinical Psychiatry

What we're supposed to think:

'Moniz's first twenty cases all survived and did not develop any serious morbidity. The leukotomy soon achieved a good reputation in, among other countries, Brazil, Italy and the United States. Moniz strongly believed that the potential benefits of surgical lesions in the frontal lobes, even allowing for some behavioral and personality deterioration, outweighed the debilitating effects of severe psychiatric illness...'
—Bengt Jansson, 'Controversial Psychosurgery Resulted in a Nobel Prize', an essay for the Nobel prize website

What you're not to say:

'Instead of offering human understanding... psychiatry has fabricated a biological and genetic explanation... to justify a massive drug assault that has taken a profound toll in terms of damaged brains and shattered lives.'
—Peter Breggin, psychiatrist and author

It was in April 1938 that an Italian researcher, Dr Ugo Cerletti, intrigued by his observations of cattle being slaughtered in abbatoirs, came up with idea of treating a 'schizophrenic' by an electric shock to the brain. After the first shock the man begged him to stop, shouting 'Not a second one! It will kill me!' Cerletti, following the style made famous by the Stanford University psychology experiments on volunteers, ignored the pleas, carried on, and afterwards recorded that the man had been 'cured'.

Acceptance of the new 'scientific' technique spread quickly through academic and medical circles, and nowhere more so than in Britain. One William Grey Walter at Bristol University even patented a special machine to do the electrocuting. For many years *Electroconvulsive Therapy* (ECT) was given without muscle relaxants or anaesthetic, and often led to injuries, including occasionally spinal fractures with their own disastrous effects. Today, a firm in Bristol can be found on the internet still advertising machines to give 'synchronous, high-intensity, well-developed, and well-generalised EEG seizure pattern... and ECT-induced seizure of high expected clinical efficacy'.

As Lisa Appignanesi put it in her book *Mad, Bad and Sad: A History of Women and the Mind Doctors from 1800 to the Present* (2008), patients 'loathed and feared the passivity, the scrambling of memory, the zombie-like condition of those who came back from treatment'. And in her novel *The Bell Jar*, Sylvia Plath has the heroine, Esther, say of ECT: 'If anyone does that to me again, I'll kill myself.' Tragically, Plath herself underwent ECT — and did kill herself.

Today, Electroconvulsive Therapy continues to be prescribed for certain unfortunate patients — not infrequently without their consent. But it has acquired a bad image. So these days, mind control is far more likely to be done by means of drugs. Drugs are a much more profitable business. By 1980 the numbers of 'patients being treated' this way was certainly in the hundreds of millions worldwide. Despite the fact that there is good reason to question diagnoses, let alone the claim of cures.

The Brain Doctors

In 2013, when the fifth edition of the much-consulted 'bible' of the modern psychiatric industry, the *Diagnostic and Statistical Manual of the American Psychiatric Association*, came out in its yearly update, it provoked an unusually strong reaction. Critics complained that the manual would lead to millions of people being unnecessarily diagnosed with psychiatric disorders. Shyness in children, temper tantrums, and depression following 'events' such as the death of a loved one were being medicalized, they alleged, as illnesses treatable with drugs. The spectre was of vested interests and a psychiatric profession in cahoots.

Just another round in the eternal war between popular and expert opinion? In the US, where today one in five children is diagnosed as having a mental illness,[1] parents were obviously voting with their feet, and the professionals could afford to shrug. Yet in the UK, unexpected support for the sceptics came from the professional body that represents Britain's 10,000 practising psychologists. In a statement the Division of Clinical Psychology called for the abandonment of psychiatric diagnosis and the development instead of alternatives which did not use the language of objective scientific fact, but rather recognized that mental health is, 'in essence, a clinical judgement based on observation and interpretation of behaviour and self-report, and thus subject to variation and bias'.

Put another way, diagnoses such as those for schizophrenia, bipolar disorder, attention deficit hyperactivity disorder, depression are not worth the paper they are written on. Or, as the statement put it, are 'of limited reliability and questionable validity'.

The ancient story of diagnosing madness is one of the most shocking and disgraceful examples—still unfolding today—of the abuse of power and responsibility by those who claim to be looking after us. The Division of Clinical Psychiatry's message, of course, is not entirely novel. Many people have their suspicions of mental health practices and practitioners, and many in the industry itself have tried to signal the scandal of 'psychosurgery'. But evidently they have failed. Today, the psychiatric assault on the most vulnerable members of the community is allowed to continue through the massive prescription of antipsychotic drugs—and even through electroconvulsive therapy.

In a disturbing account of the recent history of psychiatry called *Amputated Souls: The Psychiatric Assault on Liberty 1935–2011* (Imprint Academic 2013), Anthony James traces the modern history of psychosurgery back to a London conference in 1938, at which two researchers, C. Jacobsen and J.F. Fulton, presented their finding that when parts of the frontal lobes were removed from the chimpanzee brain, the animals ceased to be bothered by things like having their food taken off them.

---

[1]　For example, in 2007 about 9.5 percent of Americans aged 4 to 17, or about 5.4 million children, were diagnosed as having ADHD according to the US Centers for Disease Control and Prevention, and prescribed a range of 'mind control' drugs.

Figure 3. Matthew Hopkins, Witchfinder General. The medieval idea of an expert in the workings of the mind.

One of the audience of neurologists listening, Egas Moniz, a Portuguese neurosurgeon and Professor of Medicine at Lisbon, excitedly rose to his feet and proposed that similar 'frustration behaviour' in humans could likewise be cured by surgery. On his return to Portugal he began to perform lobotomies (the American term for 'leucotomies' as they are called in the UK) on psychiatric patients. Of those first twenty human experiments, Moniz records 'seven recovered and seven improved'.

In a way, that looks like mass-murder, but evidently not to medical men. Anyway, as one enthusiastic practitioner, Walter Freeman, joked later, 'there's plenty of Portuguese'. Moniz's method involved drilling holes in the skull and then severing the nerve fibres that connect the frontal lobes to the rest of the brain. The operation is of course irreversible.

Just fifteen years later, at least 50,000 people around the world had been given lobotomies. In England and Wales alone, some 10,365 psychiatric patients underwent the procedure between 1942 and 1954. A paper in the *New England Journal of Medicine*, in 1949, by J.L. Hoffman noted that:

> 'These patients are no longer distressed by their mental conflicts but also seem to have little capacity for any emotional experiences — pleasurable or otherwise. They are described by the nurses and doctors, over and over, as dull, apathetic, listless, without drive or initiative, flat, lethargic, placid and unconcerned, childlike, docile, needing pushing, passive, lacking in spontaneity, without aim or purpose, preoccupied and dependent.'

(Strange to say, one of the first people 'cured' by the procedure sought out the man who had lobotomized him and shot Moniz dead in 1955.)

In America, Dr Walter Freeman and Dr James Watt developed a quicker and cheaper way of carrying out the operation: using a rubber mallet they drove a sharp instrument (often a standard ice-pick) through the eye socket, through the bone behind it, and into the brain. James again:

> 'Strangely, the people subject to transorbital lobotomy by Walter Freeman and by others often stood the best chances of escaping the post-lobotomy changes in personality... because the neurosurgeon might fail to hit the nerve fibres altogether.

As late as 1995, *Black's Medical Dictionary*, a standard reference, says that the procedure offers a favourable response to 'chronic obsessional neuroses, anxiety states and severe chronic depression'. It adds: 'Patients are only considered for psychosurgery when they have failed to respond to routine therapies. One contra-indication is marked histrionic or anti-social personality.'

Yet at the same time, Colin Blakemore, sometime Professor of Physiology at Oxford University and Chair of the UK Medical Research Council, was prepared to describe the use of psychosurgery upon the mentally ill as 'the closest that we come to the horror of socially-applied control of the brain, and it illustrates both the inadequate restraints on the treatment of the mentally sick and the poor theoretical basis for that therapy'. However, by then psychosurgery had long moved on to two new techniques. Electroconvulsive therapy, or ECT, had become the 'convenient and quick' way of achieving much the same effect as surgery, even though there seemed to be 'no theoretical basis to justify it', as the 1987 *Oxford Companion to the Mind* puts it.

The ECT procedure artificially produces convulsions in the patient. It was developed by an Italian, Dr Ugo Cerletti, in 1938, along with Emil Kraepelin in Munich, the man whose classification of psychotic illnesses is still used today, despite having no clear theoretical basis. As James points out, the fact that both men were operating within the frame of Fascist society is by no means irrelevant—indeed Portugal was another harsh dictatorship.

Like Egas Moniz, Cerletti developed his cure through experiments on animals—on dogs and pigs particularly. He also visited slaughter houses in Rome where he observed that pigs seemed to be anaesthetized by electric shocks prior to being killed by stabbing with a knife. This seemed to him to offer a promising line of investigation for humans.

Not so long ago, in 1977, it was estimated that psychiatrists in Britain used ECT on at about 10,000 patients, typically in sequences of ten sessions per patient. And patients continue to be treated with ECT in the UK, not infrequently against their will, under the Mental Health legislation that gives doctors and social workers the power to impose 'treatment'. Refusal of treatment for mental disorders is always (as James says) a *Catch-22* situation — because specialists can assert that it is proof that the patient is unable to think properly that they do not accept the treatment. Add to which, Informed Consent is always a bit of a medical myth. Few of us will be clear-minded enough to be able to stand up to a team of medical experts, all assuring us that the procedure is 'safe', particularly when suffering from illness or severe depression.

Today, as Anthony James, the writer, observes:

> 'Despite great advances in recent years, and despite the claims of some scientists and writers — often in good faith — knowledge of the human brain is very, very partial and imperfect even today. Or rather, it would be more accurate to say that while there is now considerable knowledge of the function of individual parts of the brain, the way in which the brain works as a total system in all its varied aspects, combining thought, memory, emotion and constantly interacting with the environment, remains almost a mystery.'

The only real change in clinical psychology is that physical intervention in human minds is far, far more widespread. And that is far more likely to be done by means of drugs. As Peter Breggin puts it:

> 'The neuroleptic drugs are chemical lobotomising agents with no specific therapeutic effect on any symptoms or problems… they produce a chemical lobotomy and a chemical straitjacket.'

Dr Breggin, an American psychiatrist, is sometimes dubbed 'The Conscience of Psychiatry' for his many decades of efforts to reform the mental health field. His many books and articles have provided a foundation for modern criticism of psychiatric drugs and ECT, as well as suggesting new, more caring and effective therapies. His arguments have not been entirely welcomed, not even as debate starters. The head of the National Alliance for the Mentally Ill has publicly called him 'ignorant', and asserted that he was a publicity seeker motivated by a lust for fame and wealth. The former director of the National Institute of Mental Health branded the 'Conscience of Psychiatry' an 'outlaw'. And the president of the American Psychiatric Association says of their errant practitioner that he is the modern equivalent of a 'flat earther'.

Whatever the side effects, by 1980 the numbers of 'patients being treated' this way was certainly in the hundreds of millions worldwide.

This despite the objection (quickly made for alternative medicine) that there is good reason to question the claims of cures, claims which typically do not allow for the possibility that any improvements observed might have taken place anyway. As Peter Breggin put it in the quote opening this chapter:

'Instead of offering human understanding... psychiatry has fabricated a biological and genetic explanation... to justify a massive drug assault that has taken a profound toll in terms of damaged brains and shattered lives.'

One own particular hallmark of the chemical straitjacket is called *tardive dyskinesia*. A report by the American Psychiatric Association, also in 1980, found that 40% of patients prescribed drugs for long periods developed this illness that involves involuntary movements, most commonly affecting the lower face.

James offers a slightly rhetorical general conclusion, but surely his point is sound.

'If we have blind faith in the authority and expertise of any elite within society, political, medical or technical, we lose the capacity to think clearly... We need to rid ourselves of the notion that it is acceptable to alter the brains of certain people by chemical, electrical or surgical means, frequently without consent or on the basis of very doubtful consent. This notion is not compatible with our present scientific knowledge, and it is not compatible with conception of human dignity that has developed over the last four centuries.'

Fortunately, despite the solid front in the United States in support of current practice, there are signs that just possibly something might be beginning to change in clinical psychiatry. In what one UK newspaper excitedly called 'a groundbreaking move that has already prompted a fierce backlash from psychiatrists' (fierce backlashing!), the British Psychological Society issued a public statement formally announcing that, given the lack of evidence, it was time for, yes, it used the term, a *paradigm shift* in how the issues of mental health are understood.

The British Psychological Society statement essentially rejected the predominant biomedical model of mental distress – that people are suffering from illnesses that are treatable by doctors using drugs. And the statement carried extra weight as it emerged from the BPS's division of clinical psychology (DCP) which represents actual practitioners. The statement explained that the decision to speak out reflected these practitioners' 'fundamental concerns about the development, personal impact and core assumptions of the (diagnosis) systems' used. The leading body representing Britain's clinical psychologists now said that there was no scientific evidence that psychiatric diagnoses, such as schizophrenia and bipolar disorder, were either valid or useful.

'On the contrary,' Dr Lucy Johnstone, a consultant clinical psychologist who helped draw up the DCP's statement, wrote, there was now 'overwhelming evidence that people break down as a result of a complex mix of social and psychological circumstances — bereavement and loss, poverty and discrimination, trauma and abuse'.

Yet, if there *was* a paradigm shift taking place, it evidently had not been completed. The British psychiatrists' statement, after all, had been deliberately timed to come out shortly before the release of the fifth edition of the American Psychiatry Association's *Diagnostic and Statistical Manual of Mental Disorders*. It was the influence of this enormously influential book in creating new kinds of disorders that was being challenged.[2] And indeed, the new issue of the manual (the first for two decades) was set to add previously unsuspected new medical conditions, such as manifestations of grief, temper tantrums, and worrying about physical ill-health. These became 'major depressive disorder', 'disruptive mood dysregulation disorder', and 'somatic symptom disorder' respectively, and all of course were now clinically treatable.

Changes to long-held patterns of belief, as Thomas Kuhn predicts, are not achieved easily. Responding in newspapers to the psychiatric foot-soldiers' initiative, Professor Sir Simon Wessely, a grandee in the field as member of the Royal College of Psychiatrists and chair of psychological medicine at King's College London, said it was just wrong to suggest psychiatry was focused only on the biological causes of mental distress. In an *Observer* article he finessed the need to create classification systems for mental disorder, explaining:

> 'A classification system is like a map. And just as any map is only provisional, ready to be changed as the landscape changes, so does classification.'

But he did not discuss the dangers inherent in following an out-of-date or inaccurate map...

---

[2] British psychiatrists use an alternative manual, the *International Classification of Diseases* (ICD) published by the World Health Organization.

Chapter Four

Inexplicable Diseases

For thousands of years, doctors have profited from inexplicable diseases and the Fear Factor. They still do.

Subject: Evidence-Based Medicine

What we're supposed to think:

'Medicine [is] the science or practice of the diagnosis, treatment, and prevention of disease.'
—*Concise Oxford Dictionary*, ninth edition

What you're not to say:

'Modern medicine is not nearly as effective as most people believe. It has not been effective because medical science and service are mis-directed and society's investment is misused. At the base of this mis-direction is a false assumption about human health. Physicians, bio-chemists, and the general public assume that the body is a machine that can be protected from disease primarily by physical and chemical inter-vention. This approach, rooted in 17th century science, has led to wide-spread indifference to the influence of the primary determinants of human health-environment and personal behavior—and emphasizes the role of medical treatment, which is actually less important than either of the others. It has also resulted in the neglect of sick people whose ail-ments are not within the scope of the sort of therapy that interests the medical professions.'
—Thomas McKeown (1978)

It is not so long ago that medical experts travelled in wooden wagons painted with outrageous claims, dispensing coloured water in fancy bottles to cure various ills. Evidently, these were not real doctors but crooks and charlatans! But brightly painted wagons impressed people back then. Nowadays, doctors must train for many years and have lots of books and 'government advisories' to achieve the same effect. And modern-day coloured water is made by huge transnational companies with gleaming steel and glass-fronted laboratories. Yet, too often, neither the diagnoses nor the remedies are any better.

In fact, despite its claims to the contrary, medical science seems to be as much a creature of 'fashion' as any teenager buying into the latest fad. Take, for example, various recent breakthroughs in promoting health. Remember the fuss relating to the benefits of low-salt or low-cholesterol diets, or the entirely false consensus that had built up by the 1970s (along with that conviction that enormous flares on pant legs were 'cool') around the danger of fatty foods? Indeed, the interesting thing about many such 'breakthroughs' is that the consensus still exists, even though it never had any scientific basis.

Take the 'fat is bad' theory. This can be traced back to a single researcher, called Ancel Keys, who published a paper saying that Americans were suffering from 'an epidemic' of heart disease because their diet was more fatty than their bodies were used to after thousands of years of evolution. In 1953, Keys added additional evidence from a comparative study of the US, Japan, and four other countries. Country by country, this impressively demonstrated that a high-fat diet coincided with high rates of heart disease. It only turned out later that Keys' country-by-country comparison had also been skewed, as he'd discarded countries that did not fit his theory, such as France and Italy with their oily, fatty cuisines.

Equally unfortunate for the theory it turned out that, on closer investigation, prehistoric 'traditional diets' were not especially low-fat after all. Indeed, if the hunter-gatherers of yore had really relied on eating their prey, they would have had a lot more fat in their diet than most people do today. The reality was, as *Science* magazine pointed out, that in the most relevant period of 100 years before the supposed sharp upturn in heart disease, Americans were more likely to be consuming large amounts of fatty meat, so the epidemic followed a *reduction* in the amount of dietary fat Americans consumed—not an increase. No wonder the American Heart Association (AHA) issued a report in 1957 stating plainly that the fats-cause-heart-disease claims did not 'stand up to critical examination'.

The case for there being any such epidemic was dubious too—the obvious cause of higher rates of heart disease was that people were living longer, long enough to develop heart disease. But it was too late: the cascade had already begun. And three years later the AHA issued a new statement, reversing its view and abandoning its doubts. It had no new evidence, but it did have some new members writing the report, in the form of Keys himself and one of his friends. Their new report made the cover of *Time* magazine and was picked up by non-specialists at the US Department of Agriculture, who then asked a supporter of the theory to draw up 'health guidelines' for them.

Soon, scarcely a doctor could be found prepared to speak out against such an overwhelming 'consensus', leaving only a few specialized researchers to still protest. And all this was good enough for the highest medical officer in the US, the Surgeon General, to issue a doom-laden warning about fat in foods, insinuating that ice cream was a health menace on a par with smoking tobacco. It was the high point of a pretty feeble theory, and a triumph of preference over evidence. Only many years later, after large-scale studies in which comparable groups were put on controlled diets (low-fat and high-fat), would a correlation be found. However, it was not what was expected. It turned out that the *low-fat* diet seemed to be unhealthy. To today no one is quite sure why.

Up to then, the idea that 'fat is bad' spread far and wide—it 'cascaded' down, as social scientists put it, like water down a pyramid, from one original authority to the masses. The theory also behaved a little like a virus, infecting people by contact, particularly with their doctors.

Yet although scary in their own way, and certainly having real enough effects, 'information viruses' aren't of course the same as 'real' ones. For one thing, real ones seem to spread less effectively than we imagine they should. Take that medical mystery and seasonal misery — the common cold—for example. Back in 2003, a particularly nasty variety held the world in thrall. The condition, which was dubbed Severe Acute Respiratory Syndrome (SARS for short), included high fever and breathing difficulties, and came with an incubation period of two to seven days, which allowed those infected to unwittingly become carriers crossing continents before any symptoms appeared...

For many years there had been a growing fear among infectious disease specialists that the increase in global air travel could assist a disastrous spread of a new and lethal infection. Sniffling passengers boarding Jumbo Jets had long been a staple of plague-minded disaster movies. Infectious disease specialists often made comparisons with the flu of 1918, saying that it was the unprecedented movement of troops at the end of the First World War that precipitated the pandemic of Spanish flu across many continents. Over the course of the following year, SF is thought to have killed tens of millions of people (Spanish flu that is, not science fiction). But, back then, continental journeys took weeks—now, an infectious person can circle the globe in 24 hours.

The SARS outbreak is thought to have begun in China, but it became dramatic news after it spread to the Vietnamese capital of Hanoi via an American businessman who arrived on a plane from Shanghai and promptly infected hospital workers before himself dying of the disease. This seemed to demonstrate how dangerous the disease might be. A World Health Organization expert who bravely went to investigate the disease then themselves succumbed to it, their last few grisly hours spent in a foreign hospital bed, hastily boxed in with glass, their death adding a touch both of tragedy, and yes, human heroics to the drama.

'Fears spread over deadly virus', announced the BBC World Service, explaining that health authorities around the world were struggling to contain a lethal form of pneumonia even as air travellers spread it across the globe. The World Health Organization (WHO) issued an emergency warning, despite it being the weekend, declaring the sickness 'a worldwide health threat', adding that cases had been reported on three continents, with more 'suspected' in other parts of the world. How many sick people make a worldwide emergency? But, already, there had been four deaths related to the pneumonia and 'another five in an outbreak of a similar infection in a province of China', although the two had 'not yet been definitively' linked.

Over the next few days, more examples came in. Reuters news agency said that, in Geneva, two people were in isolation in a hospital after 'exhibiting symptoms similar to those of the disease'. In London, a British man returning from Hong Kong was taken into hospital with suspicious symptoms—providing the UK television with its first SARS case. And in Australasia, two women who had recently travelled to China were hospitalized. The Prime Minister of New Zealand, Helen Clark, confirmed the scale of the problem at a news conference and revealed that WHO experts believed the toll from the disease might rival 'Spanish flu' in 1918—which killed at least 20 million—and possibly double that number.

Back in Britain, a professor of virology, Richard Tedder, told the BBC News: 'What Sars has done is rekindle the concept of the global village.' This was by demonstrating that 'somebody's problem on a peninsula in South East Asia is Toronto's problem a few days later'. The professor urged stringent surveillance precautions at ports and airports for the foreseeable future—perhaps for years to come. Perhaps hearing his advice, the health spokesman for the parliamentary opposition in the UK demanded the arrest and quarantining of all recent travellers from Asia, a policy which would have necessitated restrictions on tens of thousands of people had it been implemented. That it never was may have been because none of the six cases of SARS at that point discovered in the UK could have been detected on arrival at the airport because none of them (due to the incubation period) had symptoms at this point.

In the event, over the whole globe, some 800 people would die with SARS, a number which, although tragic for individuals, is vanishingly small compared to the annual toll of flu and colds. The unremarkable flu of the winter of 1999 (overshadowed by the much more interesting 'Millennium Bug' that year, that is the computer problem that the relevant experts predicted would cause many of the world's computer systems to fail) killed 21,000 people in Britain alone that winter.

Undaunted, within a few months of the 'all clear' from SARS, the alert was sounded against another virus — the H5 avian variety, otherwise known as bird flu. 'We at WHO believe that the world is now in the gravest possible danger of a pandemic', said Dr Shigeru Omi, addressing a conference in Vietnam, this time adding that the world was now very 'overdue' for a pandemic, as mass epidemics usually occurred every 20 to 30 years, and it was already nearly 40 years since the last one.

A renowned virologist, Robert G. Webster, published an article in *American Scientist* magazine saying that the world was 'teetering on the edge of a pandemic that would kill a large fraction of the population'. This time, with this virus, possibly *billions* of lives were at stake. A newly appointed UN Coordinator for Avian Flu earnestly hoped measures could keep the death toll to between 5 million and 150 million. Large transnational pharmaceutical companies rushed to start preparing 'pre-pandemic' vaccines. In the US alone, $7 billion were allocated to help pay them.

By comparison, each year some 2 million people die from diarrhoea around the world, largely as a result of not having access to clean water, and the cost per person of preventing that is put at between $4 and $50 per death. The likely cost of providing sanitation for all, therefore, is less than the money cheerfully found for possible pandemics. Just one reason why public health authorities needed to justify the resources being diverted.

In Britain, these authorities predicted up to 750,000 deaths, with a minimum lower limit, graciously conceded, of 50,000. When a dead swan was found slumped on a beach in Scotland, reporters filmed (at a safe distance) health experts stumbling through the seaweed to the unfortunate beast, clad in virus-resisting spacesuits. How bad could things get, everyone wanted to know.

'At the moment the most serious concern is H5 avian influenza in chickens in south-east Asia. If this virus learns to transmit from human to human then it could sweep rapidly around the world. The 1918 influenza outbreak caused 20m deaths in just one year: more than all the people killed in the First World War. A similar outbreak now could have a perhaps more devastating impact... It is not in the interests of a virus to kill all of its hosts, so a virus is unlikely to wipe out the human race, but it could cause a serious setback for a number of years. We can never be completely prepared for what nature will do: nature is the ultimate bioterrorist.'

Or so warned Professor Maria Zambon, a 'virologist' and head of the Health Protection Agency's Influenza Laboratory in the UK. Yet even as she spoke, the virus fizzled out. Within a year the horror had passed. The worldwide toll from the disease was put at 262. Not 262 million, nor even 262 thousand, as would have been respectable, but just 262. And not one death in Britain (apart from the swan). Another health fiasco was brushed under the carpet.

Unrepentant, on 11 June 2009 the World Health Organization swept back, declaring a 'six-level alert' — its grimmest ever — for a new pandemic sweeping the world: the so-called swine flu virus. Once again, the deathly warning sounded that 'this early pandemic and flu is somewhat similar to the 1918-1919 pandemic swine flu that killed millions. What this strain will do in the Fall and Winter of 2009 and into 2010 is unclear, but everyone needs to be prepared.'

This influenza strain had only appeared around March, April 2009, but had already spread to every state in the US and throughout many places on the planet, doubtless helped by sniffling travellers on aeroplanes. The Center for Disease Control in the US started a website page to keep track of the death toll for 50 US states and territories: by June 2009 the HTML counter registered 6,506 cases and 436 deaths. The next month, reporting globally, the World Health Organization totted up on its website total cases already at 177,457 and 1,462 deaths. How many more might die though? Basing their view on US statistics, and the lack of a jab for swine flu (or H1N1, to give the virus its 'posh' name), experts thought there would be about 300 million at risk initially, 'typically, anyone who has not had the vaccine'. Both organizations pledged to keep updating their websites every week.

Bookmark these! The Center for Disease Control <http://www.cdc. gov/flu/weekly/fluactivity.htm> WHO <http://www.who.int/csr/ don/2009_08_12/en/index.html>

There was only one way to save yourself and avoid becoming a statistic: 'Get vaccinated as soon as possible.' Unfortunately, the vaccines would be available only in… early October. Meanwhile there were helpful hygiene tips though such as to:

* Wash hands regularly, before eating or drinking and after visiting the rest room.
* Avoid putting fingers and hands to the mouth or eyes since these are portals of entry for microbes.
* Stay away from large crowds, and all infected people.
* Remember to wear a face mask. Certainly if going out of the house.

Other tips to stop the spread of infections in school included teaching your kids the correct way to cover their sneezes, which, by the way, is 'sneezing into the crook of their arm'. This keeps the germs in an out-of-the-way place, preventing them from spreading everywhere. Another good tip involved 'using antibacterial wipes'. 'Sending anti-bacterial wipes to school with a child is a great way to help them avoid swine flu at school', advised the experts. Wiping desks, lunch tables, and other communal surfaces would help kill viruses and other germs, and prevent infection. Even shared pencils, pens, and the covers of schoolbooks could be cleaned with antibacterial wipes, to sanitize their surfaces.

Where necessary, the authorities promised, schools and offices would be closed. Back at home in the evening, don't forget to disinfect dishes, cups, and utensils by thoroughly soaking in detergent and washing and rinsing thoroughly everything by hand or in the dish-washer. Wash hands afterwards. And if, despite staying away from work, ordering the shopping by phone, and washing the dishes, you still fell victim—what are 'The Symptoms'? Fever, chills, coughing, fatigue, congestion, muscle and bone ache, and vomiting and intestinal upset. And then death. No wonder governments spent so much to com-bat the threat. No wonder, more specifically, that governments swiftly came up with large amounts of money to buy huge amounts of vaccine from pharmaceutical companies.

In the UK, to deal with the new threat, the impressively titled and ennobled Chief Medical Officer, Sir Liam Donaldson, warned that without a £1 billion ($1.5 billion) emergency programme of vaccination, up to 65,000 people would die. The most 'optimistic assessment' was for 19,000 deaths. His fears were confirmed by virologists. One such, Dr John Oxford, added that without immediate action soon *half* the popu-lation could be infected. Imagine, half the population dying—and only the government able to save them.

However, both Sir Liam and Dr John were immediately contra-dicted by other experts offering a very different perspective. Because one thing about experts is that they are not people who state what everyone knows, but rather people who state what only a few — the enlightened few — know. (Otherwise they're offering not expert advice but mere common knowledge...) It's a recipe for disagreement and that's how science should be. But not public policy, and here indeed the Chief Medical Officer swept aside such interventions, saying that the sceptics were extremists, and instead instructed the British government to order without further delay 32 million face masks (one between two, or perhaps he only hoped to save the small children and the ladies) and £1 billion plus worth of vaccines. In France and other European countries, a similar story was played out — another billion dollar supply of vaccines stockpiled here, another mountain of face masks there.

Yet, at the end of the day, swine flu too proved to be a paper tiger, just as the sceptical doctors had indicated. A year on, annoyingly for the governments and their advisors, almost no people could be found to be said to have died from 'HD5NI', even though flu regularly kills several tens of thousands of people each winter. However, for a while, each of these viruses was the public health concern, we could say the *fashion.* Quite possibly more people, suffering from other complaints, died from being refused entry to doctors surgeries, as part of the 'emergency precautions' surrounding the virus — or of course from being injected with the vaccine. But these could hardly be added to the statistics. And all over the (rich) world, millions of germ masks and vaccines began to deteriorate in storage, unused and unusable.

That erroneous piece of expert advice cost an unknown number of lives and enormous sums of money. Not long after, two independent reports, one by the Council of Europe and another as a paper for the *British Medical Journal* (3 June 2010), put a belated spotlight on the fact that three of the crucial experts arguing for expensive programmes of vaccine preparation by companies like Roche (the makers of Tamiflu) and GlaxoSmithKline (the makers of Relenza) were also paid consultants for the companies. Was there a conflict of interest? There were mutterings, but there seems to be no widespread resentment of the way that senior public health administrators and pharmaceutical company advisors swap places.

The money for fighting the new pandemic was not even money that would otherwise have gone on new warships or motorways but resources plucked rather from very limited budgets supposedly there for public health. Swine flu coincided with two other 'mystery diseases', MRSA and C-difficile, which killed 10,000 Britons in 2007 alone. However, there was no public emergency announced to combat these, even though the causes were fairly clear right from the outset. These diseases resulted from poor hygiene standards in Britain's public hospitals. The remedy? Spending more money on keeping them clean. But money for that, it seemed, was not there.

One British political commentator, Simon Jenkins, pointed out that no one seemed to be responsible for any errors. Looking at the flu alerts and mass vaccination programmes in 2010, he wrote:

'I am not aware of any Whitehall or Commons committee, any National Audit Office or competition inquiry into the supply of these drugs. All I know is that a huge amount of health money, time and effort was last year diverted from possibly critical therapies into what looked from the start to be yet more terror virology. This is why people are ever more sceptical of scientists. Why should they believe what 'experts' say when they can be so wrong and with such impunity? Weapons of mass destruction, lethal viruses, nuclear radiation, global warming... why should we believe a word of it?'

It's not even a bias in the rich world towards its own citizens though — it's a bias in the rich world towards rich people. In 2001, £8 billion ($12 billion) was available to stop the spread of a kind of mild cough in cows — foot and mouth disease — that affected some farms in Britain — the sole aim being to protect the value of meat exports. The money largely went on rounding up six million animals, from cows to pet dogs, and burning them in mass pyres in pits in the countryside. The army was deployed and every citizen in the UK was forbidden to enter 'the countryside' for six months.

Now, the British are famous for their love of animals. And this was not the first major animal health scare in the UK. In 1995, experts had warned that the virus for BSE/CJD in cattle would 'lead to 136,000 human deaths'. When 'senior scientists' in Britain predicted this and said that BSE/CJD 'could infect up to 10 million Britons', it swiftly led to a frenzied £5 billion campaign of cattle destruction and compensation.

Bruce Charlton, Professor of Theoretical Medicine in the UK, has an explanation though for all kinds of health madness, which he calls 'zombie science'. Zombie science is science that is dead but will not lie down.

'Zombie science is supported because it is useful propaganda. Zombie science is deployed in arenas such as political rhetoric, public admin- istration, management, public relations, marketing and the mass media generally. It persuades, it constructs taboos, it buttresses some kind of rhetorical attempt to shape mass opinion. Indeed, zombie science often comes across in the mass media as being more plausible than real science.'

Of course, there are infectious diseases amongst farm animals, especially the factory farmed ones, of course many people actually do get flu and colds, and some of them actually do die each winter, especially the poor ones, just as each year cancer, heart disease, and allergies take their toll. Of course it would be a fine thing if health experts could save us from everything, all the time. But their prescriptions come with a cost—measured in terms not only of money, but in peace of mind and in changes to lifestyle and, worst of all, in unintended side effects.

Take that evergreen recommendation to get a flu shot. Flu vaccinations are particularly emphasized for children under the age of 5, since they are 'at a heightened risk of complications from the flu virus'. But small children are also particularly vulnerable to side effects from vaccinations.

One problem is that eggs are used for making some vaccines. The virus is injected into a chicken egg, allowed to grow and multiply, and then used to create the vaccine. As a result of this, the vaccine contains tiny amounts of egg protein. But to individuals with an egg allergy, even this tiny amount can be dangerous.

Another risk factor is that a preservative, called Thimerosal, used in making many vaccines to prevent the growth of bacteria or fungi, is itself potentially dangerous, linked, it seems, to a condition called Guillain-Barré Syndrome. This sinister neurological condition is a side effect that can manifest within a day or so of vaccination. Early symptoms include numbness and weakness in the legs, which then extends to the upper body. Although the risk is reduced as cases are rare, equally the problem is increased as there is no effective treatment and it can be fatal. Patients with severe cases of this condition require hospitalization, and may need artificial respiration to assist in breathing. Because Thimerosal is mercury-based, it is nowadays avoided for most children's vaccines, but it is still used for certain multi-dose formulations. 'H1N1' or swine flu shots was one of them. Yet, as to that risk, the experts were surprisingly sanguine, saying that 'the risks of getting GBS from the H1N1 shot' were… only the same as the risks from any other vaccination.

Not everyone *was* reassured. United by health-oriented chat sites on the internet, plenty of users could be found with a rather different, and often rather pathetic, message. (Spelling as in original posts.)

> Guest :* I've seen what Gillian-Barre can do. This is not a disease that can be easily conquer and MS is almost a certainty afterward… Think about it. People need to smarten up about hygiene no vaccine is going to fix it. Oct 28, 2009 6:41 AM *

> Guest :* This is terrible!! The media has made us so afraid of NOT getting this vaccine but if you research it you become very afraid to get

the vaccine. I would love to know which risk is higher/worse… I have two 2 year olds and simply don't know what to do.

Guest :* My niece received the h1n1 vaccine at the same time as the flu shot vaccine. she is experiencing lots of swelling and hives. shes only in kindergarden. she was taken to the emergency room and they said they can not do anything. the doctor that administered the shots should know what to do. but when she is taken back to the doctor, she says there is nothing she can do. some days she experiences swelling in half of her body, or all over. nobody wants to talk to her in school anymore. she has no friends.its all just so sad.

Guest :* Last year I got both the regular flu shot and also the H1N1 vaccine. Shortly after I was diagnosed with Lupus Anticoagulant and as a result, my neurologist said I also had a mini stroke. I am being followed for the Lupus Anticoagulant, and at this time, 7 months later, the Antibody is not detectable in my blood. I don't know whether to get this year's vaccine. Doctors don't have much to say about this.

Guest :* yesterday i got the vacine for H1N1 today i can barely move every muscle in my body is sore, i have a wicked head ache and am very very thirsty… i am coughing and i have chills up and down my body. hot one minute and cold the next… how long will this last ? you left out how long the side effects last in your article…

Guest :* Everybody do your reasearch before you go and jab yourself our bodies were created to fight germs.

But not all doctors, far less patients, do 'do their research'.

The advantages of drugs are freely explained to patients, but discussion of the possible hazards is left to those experts, to take place behind closed doors. Over-prescription suits both doctors and patients alike. And so patients visiting doctors to complain of comparatively minor and 'normal' ailments, such as cold and flu, are presented with drugs about which they have no way of knowing anything about the side effects. Doctors seem, if anything, to relish prescribing increasingly powerful drugs, with increasingly obscure mechanisms.

According to a 2003 report in the journal *Paediatrics*, Americans made 25 million doctor visits in 1998 for treatment of colds. Another 1.6 million went to hospital emergency departments as out patients. And thus, every year, American doctors write millions of antibiotic pre-scriptions for colds and flu, often under pressure from worried parents. But as Snyderman says, no antibiotic, whether it's called impressively *Amoxil* or even *Zithromax*, will help a cold or flu caused by a virus. Because these drugs do only one thing: kill bacteria. And most colds and, by definition, flu are caused by viruses, a class of germs that aren't anything like bacteria.

Yet even as antibiotics are powerless to affect cold or flu viruses, they can have a powerful affect on the bodies' own natural germ-fighting system. Children taking antibiotics can suffer from stomach cramps, and other side effects.

Some experts criticize such 'over-prescription'. Grace Lee, a paediatrician at Children's Hospital in Boston who specializes in infectious diseases, says, for example, that products that promise to relieve symptoms such as stuffiness, runny noses, and coughs are ineffectual, and recommends that old standby in every medical chest of 'nothing' as the best treatment for a cold. 'Parents may want to give their children acetaminophen for a fever', she says, but for herself, if her daughter has a cold, she doesn't give her anything. And a good many studies in *Paediatrics*, the journal of the American Medical Association, *Paediatric Annals,* and other medical journals bear her out.

However, anxious parents and sufferers themselves are not reassured by nothing. So they rush to the drug store. A need for diagnosis and drugs seems a useful preliminary to any healing process taking place. At which point, consider the peculiar case of that other form of flu, Chronic Fatigue Syndrome (CFS). Well, it's called 'yuppie flu', anyway. It is also a phenomenon, like influenza, affecting millions of people (and hence potentially a huge drug market). Unlike flu though, it is not an illness that comes and goes in short episodic bouts, but rather seems to target certain individuals, often for years on end. Support groups for the illness claim that up to 1% of the population suffer from it. For sufferers even a small exertion, like washing the dishes or writing an essay, can cause a relapse lasting for days, weeks, or even months. (Curiously, the malaise may be delayed by a day or two too.)

Symptoms are most common amongst women, but the syndrome can affect anyone, including children. If CFS is often thought of as a 'new' illness, in fact it has been around since the 1930s when several epidemics of it were reported. In the 1950s it was graced with a name, when it was called ME (pronounced not 'me' of course, although sufferers were often considered to be excessively focused on themselves, but M.E.), short for 'myalgic encephalomyelitis', meaning 'inflammation of the brain and spinal cord causing muscle pain'. Nursing pioneer Florence Nightingale herself has been adopted by sufferers as a kind of emblem, as they think she too suffered from the syndrome.

However, the disease only really took off when it gathered new attention for itself in the 1980s in the United States and was 'renamed' to chronic fatigue syndrome there in 1988. The new name has been criticized for belittling the illness, as almost all chronic illnesses cause fatigue, and this one has a whole load of other alarming symptoms including:

- muscle weakness, cognitive dysfunction ('brain fog'), orthostatic hypotension (low blood pressure on standing up), headaches, fever, swollen lymph nodes, cardiac palpitations and arrhythmias, nausea and vertigo.

That's why, in the United States, some people prefer to use the term 'CFIDS' for 'chronic fatigue immune dysfunction syndrome'. But even with its long list of symptoms, other illnesses still encroached on its territory and jostled for a slice of the action. Because ultimately only one thing sets CFS/ME apart from other conditions and that is, well, feeling chronically fatigued.

Experts do not claim to know the cause. It might be bacteria – or viruses – or even chemicals. Nor do they agree on the treatment. Some recommend that sufferers try each day to take a little more exercise, but others insist on the contrary that the best way to avoid relapses is to do only *half of what you think you can do,* adding that if it takes time to accept the limitations of your body, it is crucial to avoid getting more ill.

And when neither the cause nor the treatment is known, there is plenty of room for drug companies to make good profits. For CFS there is a dazzling range of drugs, such as LDN, a prescription drug that works by increasing the secretion of endogenous opioids (endorphins) and possibly by reducing brain inflammation. It is also used for cancer, autoimmune diseases, and HIV/AIDS. It is offered to reduce sufferers' sensitivity to light and sound. Problems sleeping are sometimes treated with Baclofen (Lioresal), a muscle relaxant.

But don't despair. Curiously, even as both the cures of modern medicine and the demands of modern patients become more and more sophisticated, it turns out that many diseases respond very nicely to... cups of tea. Tea has been considered for possible powers to reduce heart disease and cancer through those famous 'antioxidants', but it is also thought to be effective over those everyday nuisances, colds and flu. In an impeccably scientific study some American researchers tried to see if special preparations of *Camellia sinensis* (green tea) capsules taken twice a day, especially with meals of course, could ward off seasonal symptoms. Not entirely to neglect helpful effects on the 'mean *in vivo* and *ex vivo* proliferative and interferon gamma responses of subjects' peripheral blood mononuclear cells to {gamma}{delta} T cell antigen stimulation', that the journal reported unintelligibly too.

Forget the jargon. In fact, drinking tea has been considered to promote good health for thousands of years, even of no one has ever proved any connection. Most studies instead have just been observational, claiming correlations between being a regular tea drinker and, for example, avoiding heart disease. In some studies, tea has failed to

show any health benefits. But then, such studies are further complicated by the nature of tea beverage. There are hundreds of varieties of
the tea plant, *Camellia sinensis*, and numerous ways to process, store,
and brew tea that can result in different components in a cup of tea, as
tea-lovers curling up their nose at brown dishwater tea in cafés know
only too well. There isn't even agreement on what quantity constitutes
a good pot of tea. No wonder, then, that attempts to evaluate the health
benefits of tea have resulted in numerous published studies all
contradicting each other.

For chemists, tea provides plenty of interesting oddities. Experts say
that there is even an amino acid, L-theanine, that is found uniquely in
tea, and then is 'catabolized' (whatever that is) to ethylamine, a
molecule that specifically activates human lymphocytes in the blood-
stream to proliferate and make interferon gamma. And here we get to
one possible respectable explanation for the long-held herbalist claims
for tea, because interferon gamma is known to have a powerful anti-
microbial effect. One study found that people who drank six cups of tea
per day had up to a 15-fold increase in interferon gamma production in
as little as one week. Coffee drinking, on the other hand, had no such
effect.

Add to which, it seems that one antioxidant found in green and
white teas, Epigallocatechin gallate (EGCG), can directly kill bacteria
and viruses, including the influenza virus, defying not only the makers
of expensive drugs but also the small area of consensus in health
matters that one preparation cannot treat both bacteria and viruses.

Based on these promising facts, some researchers supposed that
taking tea pills would have an effect, even without the evidently very
therapeutic but rather drawn-out 'tea ceremony' of china pots and
comfy chairs. And so they recently recruited some 124 healthy adults to
take the tea pills one winter. Sensibly, the adults were selected to
exclude those who already drank more than one cup of tea a day. This
screening might have been quite drastic in the UK, but not so in
America, where only 20% of Americans drink any tea at all, and those
who do almost invariably drink only one or two cups per day. (The
health benefits from tea are thought to require much higher amounts
than typical daily consumption.)

Each participant was given a bottle containing 180 capsules and was
instructed to take two every day, one in the morning and one in the
evening, preferably with meals for 12 weeks. And the results? The
guinea pigs taking the tea pills had on average just a shade under seven
days illness that winter ,while those taking 'tea-less' placebo pills had a
little over ten days illness. This represented a reduction in the number
of days ill for tea drinkers over non-tea drinkers of exactly one third,

which in terms of both statistics and drug trials was a significant difference.

Drinking tea is, in a modest way, a strategy for having 'more fruit 'n' veg.' as tea is a vegetable infusion, containing antioxidants and some other nutrients such as L-theanine. Since many people eat very few fruit and vegetables (fewer than 5% of Americans eat the nine servings per day of fruits and vegetables that health experts say they should, and any that do pass their way have usually often been 'processed to extinction'), any additional nutrients from tea are potentially significant. In due course, the tea drinking study found significant decreases both in the number of days people had symptoms, and in the number they needed medical treatment, and so the researchers thought that their work had *enormous implications* for public health. But next time you visit your doctor, don't expect to be prescribed three months worth of tea bags. Such a thing would be considered deeply unserious, verging on alternative medicine, and that, all the experts agree, is bad medicine. *Whatever the 'evidence'.*

Chapter Five

Inexplicable Cures

*The mysteries of homeopathy,
flowers, and alternative medicine*

Subject: Healing Arts

What we're supposed to think:

'The misrepresentations of history presented by Holocaust deniers and
other pseudo-historians are very similar in nature to the misrepresentations of natural science promoted by creationists and homeopaths... we
find a wide variety of movements and doctrines, such as creationism,
astrology, homeopathy, and Holocaust denialism that are in conflict
with results and methods that are generally accepted in the community
of knowledge disciplines.'
— *The Stanford Encyclopaedia of Philosophy* (Online)

What you're not to say:

'...there are effects and means of diagnosis which modern medicine
cannot repeat and for which it has no explanation... Conventional medicine looks for and tests single, "active" ingredients, but neglects the
possibility that the herb, taken in its entirety, changes the state of the
whole organism, and that it is this new state of the organism rather than
a specific part of the herbal concoction that cures the diseased organ.'
—Paul Feyerabend, *Against Method*

To understand conventional science, it helps to look at the unconventional too. This chapter looks at a range of alternative medicines, from the almost accepted, like acupuncture and chiropractics, to the 'beyond the pale' ones, like homeopathy, and asks whether it is strictly *rational* for science to be so hostile towards them.

20 January 2010, a grey, rainy day in Oxford, United Kingdom. At precisely twenty three minutes past ten in the morning (Oxford time, we may suppose), thousands of otherwise normal people attempted to commit suicide by guzzling massive doses of homeopathy pills. What motivated this desperate act? (You might have read about it the newspapers…) Were they mad? Alienated? Despairing? But no, not mad, although despairing, maybe. These were very serious people, and their purpose very public-minded. Their intention in overdosing on homeopathic remedies was to show that if pills cannot kill people, they cannot cure them either, and hence that homeopathy doesn't work and, well, that 'there's nothing in it!' Altogether, a magnificent demonstration of public adherence to the scientific method? No one died, and another blow struck against those pedlars of inexplicable cures.

Of course, it's not just indignant members of the public in England who are vexed by the continued dispensing of homeopathic 'remedies'. Invariably (nine out of ten times, one academic survey found), when the topic appears in medical journals it is accompanied by statements loudly declaiming the scientific plausibility of homeopathy. For scholarly folk, some of the articles use pretty strong language to make the point too. One paper on homeopathy (in *Respiratory Medicine*) says, for instance, that the method's use of highly diluted material 'overtly flies in the face of science' and has led it to be regarded as 'placebo therapy at best and quackery at worst'.

The clamour has led in the United Kingdom to calls to 'outlaw' homeopathy, or at least to banish it from mainstream health care services and texts. In much of Continental Europe, India, and China, by contrast, homeopathy continues to be one of the main planks of medical treatment, along with things like acupuncture and Tai Chi. Indeed, worldwide, far more people today are treated with homeopathic treatments than with conventional drugs!

The trouble with homeopathy, the thing that never ceases to enrage its critics, is that its remedies are deliberately made very, very, very weak. Vanishingly weak, in fact. They may have exciting names like 'Arsenic' or 'Snake Venom' but of active ingredients they contain, essentially, none. Not here, let alone on grey Oxford streets, the violent and unexpected patient reaction (like dying). Equally, there may be no patient reaction at all. The little 'sugar pills', as detractors call them, are swallowed and act, for all the world, only like little sugar pills. But then at least, as the medical adage goes, they succeed on the principle of 'first, do no harm'.

The 'Paradox of Homeopathic Dilution' is that some homeopathic cures would not contain a single active molecule even if you drank a dose the size of the Pacific Ocean. Doing that would assuredly make you ill too – so no wonder the subject generates constant concern among many literal-minded medical experts. Conventional medicine, like conventional thinking, assumes the strength of a remedy is directly related to the quantity of its 'active ingredient'. The more, the merrier – except when it kills you. Two aspirins are better than one. Or rather, two expensive, branded cocktails of pain-killers are better than one. (Aspirin, being out of patent protection, is frowned upon by heath professionals, despite it having almost unique health benefits, and very few side effects – taken correctly – but that's another story...) Remember, with conventional medicine, always be careful, and never exceed the stated dose.

On the other hand, the 'Watchdogs of Reason', including professionals for the 'Public Understanding of Science' and their trusted media companions, generally conclude by saying some homeopathic cures are so feeble that you'd need to eat rat poison (to get say, arsenic) or bite the head off an adder (to really get some snake venom).

Yet the point of homeopathic cures *is* to be weak. They are not supposed to cause chemical reactions. On the contrary, the origins of the approach are in the concern of its founding figure, a German doctor called Samuel Hahnemann (born 1755, died 1843, although its principles of course considerably pre-date him), at the very real, and very disastrous, chemical effects of conventional medicine two hundred years ago.

Mind you, these concerns still apply. Take drugs starting with the letter 'A', for example. Avastin, also known as bevacizumab (made by Genentech), was in its heyday the world's best selling anti-cancer drug – with sales in 2009 of $6 billion. Six billion dollars! You can get a lot of health care for that. Or at least, you might think you could. Anyway, in the US, about half of all women diagnosed with 'metastatic' breast cancer (about 15,000 people) took the drug. It cost each of them, or rather their health fund, about $90,000 dollars for a year's protection. That's a lot of money, but worse, it turned out that the drug did not actually prolong lives – but did cause numerous dangerous side effects. At the end of 2010, the US Food and Drug Administration reversed its earlier 'accelerated approval' of the drug.

In recent years several large-scale studies have found that cancer treatments produce no measurable increases in life expectancy. One study involving nearly 200,000 patients into the treatment of breast cancer by mastectomies, published in 2014 and led by Dr Allison Kurian of Stanford University, found that people who had a double mastectomy (i.e. one being 'preventative', and one in response to a suspicious lump) had no better prospects (or worse) than those who declined the treatment—while those who opted for the conventional treatment of addressing just the lump actually did worst of all.

Cancer is a particularly strange 'disease'—because many cancers are benign, and many others cure themselves. That's the job of the body's immune system. It quietly cures cancers all the time. Or rather, nearly all the time… There's the rub. And it's hard to recommend a 'wait and see' approach, as malignant cancers certainly can't be surgically treated late as they have spread around the body.

You don't hear so much about the body's own defences against cancer though. Professor Robert van der Bosch, late of the University of California, wrote a book he called *The Cancer Conspiracy*, in which he argued that there was a kind of corporate logic to presenting cancer as a rare illness which required treatment, rather than as a ubiquitous everyday phenomenon that is best addressed through preventative measures, healthy lifestyles, and high environmental standards. He even suggests that: 'an effective form of cancer prevention would have devastating consequences for the cancer industry. Hence the emphasis is not on cancer prevention, but on treatment and diagnostic procedures, because that is where the big money is.'

And indeed, forty years of the fight against breast cancer by the billion dollar charities seem to have produced only much higher numbers of women with the disorder—a veritable epidemic. Breast cancer is today said to touch one in eight women in the rich world. Detection, chemotherapy, radiation treatment, surgery are the services on offer. But very little talk about prevention…

Another illness we are all likely to contract is the common cold. Yet, as mentioned in Chapter 4, 'Inexplicable Diseases', antibiotics cannot kill the viruses that cause colds and flu. Doctors write many millions of antibiotic prescriptions, yet none of them can cure, or even help,[1] a cold or flu. Because antibiotic drugs do only one thing: kill bacteria. Yet even as antibiotics are powerless to affect cold or flu viruses, they can have a powerful effect on the bodies own natural germ-fighting system, and cause side effects such as stomach cramps—or worse.

---

[1] They can help colds or flu that have developed complications involving bacteria, though. Don't disbelieve your doctor automatically!

Flower Remedies

A relatively recent exponent of the medical use for flowers is the British doctor called Edward Bach. His approach is similar to homeopathy's in many respects. Bach (1886–1936) was born in a village called Moseley, near Birmingham, England, and studied medicine at University College Hospital, London, before working as a practising doctor there. He then went into mainstream medicine and worked elsewhere in London as a bacteriologist, eventually joining the laboratories of the Royal London Homeopathic Hospital in 1919. He came to believe that illness was the result of a pre-existing disharmony between body and mind. Symptoms of an illness were but the external expression of negative emotional states. Armed with this theory, in 1928 he began work on his own remedies made from plants.

There are 37 original Bach remedies, using 37 fairly ordinary plants:

The so-called 'twelve healers' are Agrimony, Century, Cerato, Chicory, Clematis, Gentian, Impatiens, Mimulus, Rock Rose, Scleranthus, Vervain, and Water Violet. Gorse, Heather, Oak, Olive, Rock Water, Vine, and Wild Oat are called affectionately the 'seven helpers'.

Dr Bach also recommends potions made from 'the second 19', that is: Aspen, Beech, Cherry Plum, Chestnut Bud, Crab Apple, Elm, Holly, Honeysuckle, Hornbeam, Larch, Mustard, Pine, Red Chestnut, Star of Bethlehem, Sweet Chestnut, Walnut, White Chestnut, Wild Rose, and Willow. All of these grow in Britain, where Bach lived.

Bach flower essences are not based on any chemical theory, whether New Age or solidly traditional. Another difference between homeopathy and Bach flower remedies are the methods used to produce them: the production of Bach flower essences is handled in two ways, either drying flowers in the sun, or boiling them. And a final rather significant difference is that homeopathy uses either water or ethanol as a medium for its mysterious cures, but flower remedies use a base of good quality brandy.

Anyway, on to the search for remedies. Consider (sceptically) this one for treating 'Excessive Worries about Other People'. Excessive worries about other people? That seems a funny sort of malady, especially in an era of rampant individualism. You wouldn't think cures for it would sell very well. But these are alternative medicines, and it is well known that 'hippies bearing spirit trees' are apt to worry too much about world problems. In any case, for the method's adherents, such strange worries lie behind many conventional illnesses. And it seems that the dark-pink flower of the Red Chestnut tree, a tree which grows to some sixty feet (twenty metres) and has deeply furrowed bark, is able to dispel such excessive worries about other people.

If that doesn't clear it up though, another appropriate remedy might be Vervain, or *Verbena officinalis*. This is a flower whose particular quality is that it is thought to be able to relieve intense over-enthusiasm, something conventional medicine is slow even to diagnose. Vervain, unlike Red Chestnut, is an ancient herb with small, purply-pink flowers, that grows up to 3 feet and prefers dry, limey soils. Indeed, it has been used in herbal medicine for centuries if not longer to treat nervous problems and insomnia. It has one suspected side effect, but it is not too serious—it is also thought to provide 'inspiration' and in the past has been associated with poets and writers, if not many newspaper columnists.

Maybe over concern for others and the like don't even sound like real illnesses. But there are also people who suffer from fully-fledged FEAR, people who are terrified, and panic. For this, Dr Bach offers Rock Rose, a radiant 5-petaled flower whose blooms lasts just one day. Published, scientific research on the method is, shall we say, limited. Everyone seems to assume it is nonsense. And anyway, the Bach flower remedies are not herbal medicine as such, despite their plant origins. For example, the root of that everyday weed, Dandelion, can be made into a bitter tea (don't sweeten it, the bitterness is how it works) that is said to be an excellent tonic for the liver and a blood purifier. And it is true that Dandelion roots are very high in vitamins and minerals.[2]

Dr Bach believed that the flower essences interact with the subtle mechanisms of the human mind, and so can help people with physical illness by addressing the emotional responses to their illness. His idea was that the human being incorporates a body of life energy, a body of sensitivity and feelings, and a spiritual body. Flower essences are helpful imprints of the life force of plants.

But most herbal remedies are content to offer associations. The rose is associated with the heart, and the humble onion with the universe (on account of its layers of concentric spheres). The night blooming flowers of jasmine are considered the greatest aphrodesiac of them associated with the compassionate Goddess of the Moon. In ancient Egypt, where the sweet white flower represented Isis, it was also accorded special powers of fertility, magic... and healing.

Nonetheless, many of us fell better when we take our cold remedies. And equally, today there are still some chemists who will admit to being puzzled by the very real evidence of the little sugar pills' 'inexplicable' curative effects, and who wonder whether perhaps rejecting the homeopathic approach wholesale risks 'throwing out the baby with the bathwater', so to speak. Some speculate that our 'understanding' of the properties of water, and hence that of homeopathic pills, might be based on false assumptions. One splendidly 'modern' theory is that water may even have peculiar quantum mechanical properties the understanding of which could be a key to unlocking the secrets of life.

---

2   The common Dandelion is an unusually nutritious food. Its leaves contain substantial levels of vitamins A, C, D, and B complex as well as iron, magnesium, zinc, potassium, manganese, copper, choline, calcium, boron, and silicon. We're much happier nowadays explaining the curative effects of plants in such straightforward terms. If herbalists also add that the much despised Dandelion contains two active constituents called eudesmanolide and germacranolide, and that these substances are unique to this herb, then we more readily accept it as a 'proper medicine'.

The people who assert the impossibility of homeopathic cures over-state the state of knowledge in contemporary science—as scientists have done for thousands of years. Brian Josephson, not an alternative medicine guru but rather a Nobel laureate of physics, has argued against such cozy assumptions. Commenting on such zealous debunking exercises in the *New Scientist* magazine (which also recently debunked a book of mine making the case against nuclear power, by the way!), he notes that generally:

> '...criticisms (of homeopathy) centred around the vanishingly small number of solute molecules present in a solution after it has been repeatedly diluted are beside the point, since advocates of homeopathic remedies attribute their effects not to molecules present in the water, but to modifications of the water's structure. Simple-minded analysis may suggest that water, being a fluid, cannot have a structure of the kind that such a picture would demand. But cases such as that of liquid crystals, which while flowing like an ordinary fluid can maintain an ordered structure over macroscopic distances, show the limitations of such ways of thinking.'

Put another way, although hydrogen bonds have only a fleeting existence in liquid water (about a picosecond) and not much longer in ice, it is known that an ice sculpture (and this can easily be demonstrated) can somehow 'remember' its shape—over extended periods. Could there after all be some unknown physical mechanisms that explain homeopathic medicine?

'Mysterious forces' get short shrift in hospitals, but in fact 'natural philosophers' and scientists alike do often allow certain mysterious forces to operate in their theories, when it suits them, resting content to give them new names like 'Dark Energy' (which it seems is needed in order to hold the entire universe together). Not, however, that that means they feel comfortable with them.

Step forward, then, Madelaine Ennis, a pharmacologist at Queen's University, Belfast—and virulent critic of homeopathy. Dr Ennis regularly railed against its practitioners' claims that remedies could be diluted to the point where a sample was unlikely to contain a molecule of anything except water—and yet still have a healing effect. Naturally, she was pleased to be invited to participate in experiments designed, essentially, to prove, once and for all, that homeopathy was bunk.

Like a good scientist, Dr Ennis and the other researchers gave some patients dilute solutions of a drug (histamine, considered to boost the body's immune system, amongst other things) while, like a good homeopath, she gave other patients ultra-dilute remedies—so dilute, indeed, that they probably didn't contain a single active molecule—and waited to see the ultra-dilute treatments fail. Instead, however, she observed that, in practice, the homeopathic solution worked on the human body every bit as well as the conventional one. The study, later replicated by other researchers, forced Ennis to consider that maybe something was going on after all with the supposed pseudoscience. 'We are', she says, 'unable to explain our findings and are reporting them to encourage others to investigate...'

This preparedness to consider the inexplicable is alas increasingly rare amongst 'professional scientists' who are often dependent on the good opinions of their peers (not to mention their employers) and fearful of straying far from the path of orthodoxy. Gone are the days when people experimented freely. Darwin, for example, despite having his own controversial theory to establish, also discussed all sorts of curious and inexplicable phenomena, and did so without being thrown out of the scientific fold. Amongst his diversions, an historical curiosity — or is it confirmation of the importance of the homeopathic principle? — are the results of a series of experiments he made to examine the effects of ultra-dilute solutions on carnivorous plants. In a book called *Insectivorous Plants* (1875) Darwin recalls how his results defied his expectations:

> 'The reader will best realise this degree of dilution by remembering that 5,000 ounces would more than fill a thirty-one gallon cask [barrel]; and that to this large body of water one grain of the salt was added; only half a drachm, or thirty minims, of the solution being poured over a leaf. Yet this amount sufficed to cause the inflection of almost every tentacle, and often the blade of the leaf... My results were for a long time incredible, even to myself, and I anxiously sought for every source of error... The observations were repeated during several years. Two of my sons, who were as incredulous as myself, compared several lots of leaves simultaneously immersed in the weaker solutions and in water, and declared that there could be no doubt about the difference in their appearance... In fact every time that we perceive an odour, we have evidence that infinitely smaller particles act on our nerves.'

But we have to be careful; homeopathy was not the declared, explicit, subject of this text, although it may have been an underlying riddle for Darwin, who is known to have visited a homeopath. (And, yes, he is said to have felt better after. Evidence!)

So why the rancour over homeopathy, the shrill denunciations and vehement attacks on its defenders (let alone its practitioners)? But to make sense of many of the debates in medicine, it helps to have an historical perspective. Because, truly the history of homeopathy is the history of medical science.

So step into the Time Machine now, please, as over-enthusiastic historians say, and travel back with me to the dawn of 'scientific' medicine — which is not so long ago, being in the late eighteenth century.

This was the time of Samuel Hahnemann, the 'inventor' of homeo-pathy, when doctors were always looking for one overarching theory to explain how both the body and Nature herself worked. Not that this approach was itself exactly new.

For philosophers, one theory to explain everything is always the aim. It thus seemed to doctors eminently sensible that all diseases should be explained using the same theory. Alas, rival grand theories were the subject of much controversy, and even the conventional wisdom of one year could become 'mere quackery' and fall out of favour the next. Of course, medicine is still like this! But, in the eighteenth century, the top all-explaining theory of health was that most illnesses resulted from gastric impurities, especially bile. The removal of these matters by 'emetics' (things to make you throw up) and purgatives was the key to curing illness. If signs of bile were absent, it was a case of 'latent' bile—*bilis latens*, to grace it with a spuriously authoritative Latin name (like many medical phenomena).

At its high point, this simple doctrine was regarded as one of the most brilliant advances in the medical art. Doctors betook themselves to Vienna from all parts of Europe to learn the method of its discoverer, the celebrated Dr Stoll. Once there, they heard how diseases arose from the influence of a predominant constitution which was determined 'by the prevailing weather and epidemic fevers'. One doctor of the period wrote: 'Stoll is the greatest living physician. He stands, as he deserves, in a position of great repute, and all intelligent persons in Vienna are attended by him.'

But brilliant though he wasn't, Stoll had his rivals. Before long, another theory, that of one Dr Kampf (1726–1789), swept Stoll's aside. The new orthodoxy became instead that most diseases had their seat in the abdomen, and were due to what Dr Kampf called *infarctus*. 'By infarctus', Kampf explained, 'I understand an unnatural condition of the blood vessels.' It was a condition resulting in 'coagulated blood tarrying', that is flowing more slowly… and eventually stopping. This 'inspissated serum in the blood, in the glands, in the cellular tissue, together with the above-mentioned blood-dregs, collects, corrupts, dries, and takes on various forms of degeneration in the digestive passages'.

Dr Kampf assured his disciples that 'infarctus' spares 'no age, sex, or temperament' and that even tiny infants were not free from them. He offered up many examples of 'typical infarctus', including epilepsy, cataracts, deafness, consumption, abdominal diseases, bladder affections, cancer, scurvy, fever, dropsy, jaundice… and more. Since all illnesses had the same cause, one treatment method was enough. Dr Kampf's method, not unlike those of the homeopath doctors later, was to make a solution of rain or lime-water containing particular extracts considered to have special properties relating to the disease. These remedies, which he called 'Clysters', contained hints of traditional plant remedies, such as millefol., chamomill., and rye-and-wheat-bran; to which various 'appropriate' drugs made from chemicals were added, the whole being made into a drinkable solution.

Clysters were considered to be without side effects. Hence they could be taken in large quantities. One enthusiastic physician of the time wrote: 'I have treated many sick persons who have taken more than five thousand clysters before they entirely got rid of the *infarctus*.' Naturally, clysters were not free… and five thousand of them five thousand times less so.

This, then, was the background to the work of Samuel Hahnemann. Yet if Dr Kampf's clysters look very like Dr Hahnemann's tubes of pills in water (or ethanol), there are differences. For one, Hahnemann recommends that different approaches be adopted if no improvement is observed. He adds that 'The intellect alone can (a priori) evolve from itself alone no conception of the essential nature of things, of cause and effect; there must always be sensible perceptions for every one of its dicta concerning the actual. Facts and experience must be at the root of all revelations of truth.'

If today's critics of homeopathy often first appeal to the evidence from standardized tests of homeopathic remedies as a way to marginalize them—in fact, Hahnemann's real influence on medicine was an emphasis on testing. In the preface to the second edition of the *Organon* (his book on the homeopathic method) he writes:

> 'True medicine is from its very nature a pure science of experience, and should therefore rest only upon pure facts, and the sensible phenomena belonging to its sphere of action, for all the subjects with which it is concerned are distinctly and sufficiently indicated to its sensible appreciation by experience; knowledge of the disease to be treated and of the action of drugs and also the mode in which the ascertained actions of medicines are to be used in curing diseases, can only be learnt by experience; its subjects can only be derived from pure experience and observation, and our science should not venture a single step beyond the sphere of pure carefully observed experience and experiment if it wishes to escape degenerating into mere jugglery and nullity.'

Hahnemann only gradually worked towards the tiny—some would say non-existent—doses characteristic of homeopathy. In 1791, he speaks of giving narcotic vegetable medicines only 'in very small doses'. In 1793 he urged that when arsenic was prescribed (as it often was for ulcers or even fevers) that the usual dose be reduced by three-quarters of the then 'going rate'.

In 1805, in a book called *Medicine of Experience,* he says: 'None but the careful observer can have any idea of the height to which the sensitiveness of the human body to medicines is increased in disease. It transcends all belief when the disease has attained a great intensity...'

When a teaspoonful of tincture of *helleborus niger* twice a day is recommended by one health expert, Hahnemann remarks: 'This enormous dose should assuredly be diminished to a twentieth part. Two drops of the strong properly prepared tincture of black hellebore are enough to act powerfully on an adult, and will do all that is possible to be done in cases where the tincture is indicated, and if it is not indicated so large a dose will cause irreparable damage.'

On the face of it, today's representatives of what is called 'evidence-based medicine' should count Hahnemann as one of their inspirations. Yet instead they show him unrelenting hostility. But then medicine can be a very rough game. As a Professor Roose puts it, writing at the dawn of homeopathy, so to speak, in 1803, there exists amongst doctors, alongside a strange lack of curiosity, 'a savage partisan spirit' which regularly splits the medical profession into sects, 'every one of which embitters the others by violent and often unfounded contradiction, and so prevents all possibility of doing good.'

Of course, that was back then. Things are much worse now! Take for example the writings of one of our best known public health experts, the English popular author and newspaper health guru, Ben Goldacre. Writing in his regular column for the otherwise generally rather liberal London paper *The Guardian,* under the heading 'What's wrong with homeopathy', Dr Goldacre trots briskly through all the arguments against what homeopathy isn't. Naturally, he points out that the solutions are very weak. As if the homeopaths hadn't noticed! Then, he demands that 'the healing powers of homeopathic pills... be tested like every other pill'. Yet, it is homeopathic practitioners who pioneered the notion of careful testing.

Next, Dr Goldacre remarks that the pills may not be doing anything but the body just starts to recover anyway, and the pill is wrongly given 'the credit' as it were. Of course this is both quite likely and a criticism that applies equally if not more so to conventional medicine. To conclude, Goldacre offers, by way of a final flourish, the pheno-menon of 'regression to the mean'. 'This is an even more fascinating phenomenon', he remarks caustically: 'all things, as the new-agers like to say, have a natural cycle. Your back pain goes up and down over a week, or a month, or a year. Your mood rises and falls. That weird lump in your wrist comes and goes. You get a cold; it gets better.'

Spookily, as if somehow foreseeing the likes of Ben Goldacre centuries later, who trades heavily on his title,[3] Hahnemann adds, caustically, 'A qualified doctor is, of course, at liberty to give anything he likes, Nature must submit out of respect for his diploma.' Certainly, Hahnemann's method, whether advisable or not, had nothing to do with pundits' learned points about misreading statistics. His solution to the problem of establishing the effects of his potions was simple. He tested each ingredient out in turn upon himself, carefully noting down the effects it had. His first experiment, or 'proving' as he called it, was of cinchona bark, a cure then frequently prescribed for curing fevers, notably malarial ones.

'For the sake of experiment, I took for several days four drachma of good cinchona bark twice a day; my feet, finger tips, &c., first grew cold, I became exhausted and sleepy; then my heart began to palpitate, my pulse became hard and rapid; I had intolerable anxiety, trembling (but not rigour), prostration in all my limbs; then throbbing in the head, flushing of the cheeks, thirst, and in short all the ordinary symptoms of intermittent fever12 appeared one after another, but without actual febrile rigour. In a word, even the special characteristic symptoms of intermittent fever, dullness of the senses, a kind of stiffness of all the joints, and in particular the disagreeable numb sensation which seemed to be located in the periosteal covering of all the bones of the body, made their appearance. This paroxysm lasted two to three hours each time and returned when I repeated the dose, otherwise not. On leaving off the drug I was soon quite well.'

This showed to Hahnemann that:

1. the ingredient did indeed have an effect,
2. what it was, and
3. that it might well be dangerous.

It also seemed to confirm to him his suspicion that the ancient principle of treating diseases with agents considered to induce the opposite effect (for example, treating constipation by purgatives, chronic pain with the painkiller opium) was wrong. 'And although the great majority of my medical brethren still adhere to this method, I do not fear to call it palliative, injurious and destructive.' And so Hahnemann instead advanced the reverse principle — which is that disease should be treated with agents that induced the same symptoms. If this sounds perverse, it is of course the principle behind vaccination. With vaccines, a small dose of a disease is given to a healthy person, in order to stimulate the

---

3   The official NHS register of UK doctors notes that Ben Goldacre is not on either the Specialist or the GP Registers.

body's natural defences so that they can cope in the event of infection by the disease proper later. Another controversial figure in the history of medicine, Paracelus, had earlier popularized the approach, starting with plague germs in bread.

The reality is that, despite this, or perhaps this explains his evil reputation for esoteric 'quackery' in medical circles, Hahnemann was a very practical man, who counselled, and popularized, many simple preventative strategies for good health. He was a great advocate of exercise and open air, and also of the beneficial action of change of climate and residence at the seaside, all things which were then rarely mentioned in medical works. He warned that, without such things, people (he mentions particularly women) are 'reduced to the condition of colourless plants grown in a cellar'.

Hahnemann was also an early advocate of the importance of hygiene. Similarly, he was far ahead of his contemporaries in the treatment of mental illness. Physicians then treated excitable and refractory maniacal patients like wild animals; it was thought necessary to cow and terrify them. Corporal chastisement and nauseating medicines were the usual means used. 'Maniacs' were strapped down on a horizontal board which could be quickly turned on an axis to a vertical position, or put in the so-called rotating chair. 'A well fitted up madhouse was, in certain respects, not unlike a torture-chamber', as one commentator, Carl Friedrich Otto Westphal, put it, writing a generation or so later.

The Father of Homeopathy's treatment of the insane instead was more enlightened: 'I never allow an insane person to be punished either by blows or any other kind of corporal chastisement, because there is no punishment where there is no responsibility, and because these sufferers deserve only pity and are always rendered worse by such rough treatment and never improved.'

Indeed, early in his career, he treated and cured in this way the Chancellery Secretary Klockenbring of Hanover, 'a man well known to literature', who had become deranged. After his complete cure from madness this sufferer showed his deliverer, 'often with tears in his eyes, the marks of the blow and stripes his former keepers had employed to keep him in order'. As one biographer puts it, no physician since Paracelsus had dared to expose with such frankness and boldness the miserable condition of the medical treatment of the period.

Hahnemann was always at war with the conventional regimes of his times. As early as 1786 he had inveighed in a book against 'that most fruitful cause of death, the bungling of physicians'. In an anonymous article, in the year 1808, after he had for twenty years past been calling the attention of his contemporary physicians to recognize the alarming consequences of deficiencies in the 'healing arts', he writes:

> 'It must some time or other be loudly and publicly said, so let it now be boldly and frankly said before the whole world, that our art requires a thorough reform from top to bottom... What should not be done is done, and what is essential is utterly neglected... No other science or art, or even handicraft, has advanced so little with the progress of time, no art is so behindhand in its radical imperfection as the medical art.'

Which brings us to one small victory for homeopathy in this sense, that even the dedicated debunker Dr Goldacre himself acknowledges. It concerns the well-known story of the nineteenth-century cholera epidemic in London, a time when death rates at the London Homeopathic Hospital were three times lower than at the 'conventional' Middlesex Hospital. Three times! Surely that was a significant success for the 'sugar pills'? But no, Goldacre instead insists, on behalf of modern medicine, that the reason for homeopathy's success was that at the time the conventional treatments, such as blood-letting, were so bad that they actively harmed the patients, whilst the homeopaths' treatments at least did nothing either way.

Having made that small (he thinks) concession to alternative medicine in the nineteenth century, Ben Goldacre then gleefully recounts some dreadful sounding stories of the harm that homeopathic practitioners today do. Most impressive of all, he thinks, was the outcome of a dramatic July 2006 investigation by a BBC undercover team into London's homeopathic clinics. We should consider it too, as it says something about how scientific debates are joined and won by the media. Goldacre's words again:

> 'A BBC *Newsnight* investigation found that almost all the homeopaths approached recommended ineffective homeopathic pills to protect against malaria, and advised against medical malaria prophylactics, while not even giving basic advice on bite prevention. Very holistic. Very complementary. Any action against the homeopaths concerned? None. And in the extreme, when they're not undermining public-health campaigns and leaving their patients exposed to fatal diseases, homeopaths who are not medically qualified can miss fatal diagnoses, or actively disregard them, telling their patients grandly to stop their inhalers, and throw away their heart pills.'

Oh dear me! But given that this is a battle for the principle of evidence must trump private preferences, it should at least be noted that the research methodology of the *Newsnight* investigation required the use of an undercover 'patient' who, as the BBC health correspondent admitted, told the clinics that:

'...she did not like the side effects of the drugs the doctors prescribe, and asked if there was anything else she could take.'

That the alternative health gurus offered their remedies is perhaps less than amazing. One might even say, less than newsworthy. As for the claims about heart pills and inhalers, neither the *Newsnight* investigation nor Dr Goldacre offers further details. Possibly they are just hearsay — originating from articles and comments like these ones on a news website.

First of all, here's Jeanette Winterson, a well-known writer describing her resort to the wrong sort of doctors in an article for the *Daily Mail*, a middle-brow London newspaper with a soft spot for alternative medicine:

'Picture this. I am staying in a remote cottage in Cornwall without a car. I have a temperature of 102, spots on my throat, delirium, and a book to finish writing. My desperate publisher suggests I call Hilary Fairclough, a homeopath who has practices in London and Penzance. She sends round a remedy called Lachesis, made from snake venom. [So that's what it's for!] Four hours later I have no symptoms whatsoever.'

Disgraceful! Irresponsible! No attempt by Hilary Fairclough to refer Jeanette to her doctor... And now look at these comments the story produced from readers...

This one from 'Sue' in London:

'My husband was given antibiotics for a badly infected finger. He took them for 5 days and it got better. The next day he had a letter through the post saying that the swab results showed it was a different bacteria from the one the doctor had assumed and that the antibiotics he had been given were the wrong ones and would not work and that a prescription for the right antibiotic was waiting for him at the surgery! The finger had got better by itself!'

And finally, Elainejones, Leeds, W. Yorks, who says:

'My husband suffered from terrible migraines for years with no help from doctors; he was sceptical about homeopathy but in the end went along. He has had hardly any migraines now for 3 years and when he does the remedies always help him.'

This is the kind of hearsay that enrages dedicated rationalists. But at least Dr Goldacre can stand up for reason:

'By pushing their product relentlessly with this scientific flim-flam, homeopaths undermine the public understanding of what it means to have an evidence base for a treatment. Worst of all, they do this at the very time when academics are working harder than ever to engage the public in a genuine collective ownership and understanding of clinical research.'

That virtuous process is exactly what Dr Goldacre sees his own task as being, by the way. So, perhaps wary of the risk of sounding otherwise rather self-serving, he continues:

> 'There are also more concrete harms. It's routine marketing practice for homeopaths to denigrate mainstream medicine. There's a simple commercial reason for this: survey data show that a disappointing experience with mainstream medicine is almost the only factor that regularly correlates with choosing alternative therapies.'

Some might wonder how significant homeopaths 'denigrating' conventional medicine might be, if the only people they see are those who have turned to them because they are dissatisfied with what the conventional doctors offer. But Goldacre has a 'horror story' about that:

> 'One study found that more than half of all the homeopaths approached advised patients against the MMR vaccine for their children, acting irresponsibly on what will quite probably come to be known as the media's MMR hoax. How did the alternative therapy world deal with this concerning finding, that so many among them were quietly undermining the vaccination schedule?'

Once again it seems that homeopaths prefer their remedies to the conventional ones! And certainly medical advisors should avoid either actually having, or even the *appearance* of having, 'conflicts of interest' like this. Or, perhaps, like Goldacre himself, who studied in departments funded by the major drug companies. And then there's his receipt, in 2003, his first year as *Guardian* writer on health issues, of the prestigious British Science Writers award. At this time, the award was funded by the drug manufacturers Glaxo Wellcome who were, amongst other things, the manufacturers of the MMR vaccine. It begins to look a bit like a case of the pot calling the kettle black. But let the voice of neutral, *rational* science have the final word on homeopathy.

> 'When I'm feeling generous, I think: homeopathy could have value as placebo, on the NHS even, although there are ethical considerations, and these serious cultural side-effects to be addressed. But when they're suing people instead of arguing with them, telling people not to take their medical treatments, killing patients, running conferences on HIV fantasies, undermining the public's understanding of evidence and, crucially, showing absolutely no sign of ever being able to engage in a sensible conversation about the perfectly simple ethical and cultural problems that their practice faces, I think: these people are just morons.'

Figure 4. Try this! There is a deeply rooted tradition in the healing arts of hoping cures will work without really needing to understand the mechanism. Chinese lore allows that merely to hang a picture of Zhang Guo Lao in the house will bring good health and a long life.

Part II

Chapter Six

Physics' Guilty Secrets

and its favourite fairytales, featuring the True Story of the famous battle between the followers of the Sun and of the Earth

Subject: Cosmology

What we're supposed to think:

'Although Copernicus claimed his work was no more than hypothetical, eventually the weight of evidence would be too great to be resisted, and before long Copernicus would be famously supported by Galileo Galilei, Johannes Kepler and Isaac Newton, amongst others.'
— Philip Stokes, *100 Essential Thinkers* (Arcturus 2006)

What you're not to say:

'Not only are facts and theories in constant disharmony, they are never as neatly separated as everyone makes them out to be... Copernican hypotheses failed the evidence and ran counter to almost every methodological rule that one might care to think of today.'
— Paul Feyerabend, *Against Method*

The chances are, if someone is being sceptical about one or other vaguely scientific claim, sooner or later someone else will point them back to Copernicus and Galileo and remind them of the lesson of what happens whenever political opinion and ideology collide with the unbending rock of scientific method.

For many people the ripping yarn of the plucky Polish astronomer, whose theory that the Earth went around the Sun and not vice versa, is surely the 'paradigm case' of paradigm shifts, so to speak. Indeed, the Copernican Revolution is often said to mark the beginning of the 'Age of Science' — and to herald the start of the 'Enlightenment', as it is comfortably called.

It's a ripping yarn in which the dashing Galileo, flourishing his new-fangled telescope, battles valiantly against a Church hierarchy still clinging to its old-fangled Bible's words that: 'The sun also ariseth, and the sun goeth down, and hasteth to his place where he arose' (*Ecclesiastes*) — while relying on the sinister threats of the Papal Inquisition to keep mankind in darkness and ignorance.

Just a shame then that the facts are all quite the other way round. It's not that the Earth really is the centre of the universe, though it might be for all the difference it makes, just as Milton Keynes is the centre of England, and Kansas is the centre of the United States, but rather that in this debate the patient exponents of scientific method were on the losing side and Copernicus's triumph was achieved, like so many giant leaps forward in science, only by spin and propaganda.

But that's not what you'll read in the textbooks. Instead, here it is the wise scientists who patiently and courageously challenge the Church authorities with their foolish insistence that the Earth must be set like a jewel in pride of place at the centre of all creation while the heavens rotate around it on crystal spheres. In the official history, careful collection of observational evidence that the Earth was but a large rock rotating around the Sun eventually won the day, but alas it was a victory that came only after Galileo's conviction for heresy and imprisonment.

It's a great tale to frighten the sceptics with, even if, like a Chinese whisper repeated too many times, in too many standard texts it bears almost no relationship to the actual events. However, this is not a textbook and we can afford to be a little contrarian, so let's put aside the authorized version for a moment and instead go back to the vexed state of astronomical events in the seventeenth century.

This was a time when rulers like the King of Denmark were prepared to provide huge sums of money for scientists like Tycho Brahe to aid their efforts to draw up a detailed map of the heavens. Actually, the King's largesse was not for the love of pure research as such but rather in the hope of improving the accuracy of astrological forecasts, and hence gain a march on his enemies. Equally, if this was a period when the great 'natural philosophers', savants like Isaac Newton and Gottfried Leibniz, experimented freely, it was also a time when those investigations were mainly in the dark arts of alchemy — conducted in great secrecy and leaning heavily on ancient folklore. Not here the tidy division of the world into rational thinkers seeking to spread knowledge and irrational bigots seeking to suppress it that today's science books offer.

Even the relationship between Tycho and the King was not beyond the influence of the irrational. Tycho was a heavy drinker and fell out with his employer. It was in a rage that he took his instruments off to Germany and it was only there, after his death, that his assistant Johannes Kepler finished the map of the heavens. In order to complete his new, mathematical model of the universe, Kepler resurrected some small parts of Copernicus's earlier work. Notably, the bit where the Earth is set in motion around the Sun.

How much of a scientific revolution is that? Rather less than it might seem.

The soundbite version is that Copernicus 'discovered' that the Earth revolves around the Sun. Whereas, actually, the idea is an old one, hotly debated by the Ancients, appearing in influential Sanskrit, Arabian, and Roman scripts, and particularly well rehearsed by the ancient Greeks. Some of the accounts were pretty detailed too. In one ancient text, called 'The Sand Reckoner', Archimedes, the celebrated inventor and mathematician, records enthusiastically that someone else called 'Aristarchus' had just brought out a book on the workings of the heavens including exciting hypotheses such as that:

• the stars and the Sun remain unmoved, while the Earth turns on its axis;
• the Earth revolves about the Sun in the circumference of a circle, with the Sun lying in the middle of the orbit, and that
• the distance to the sphere of the fixed stars, starting with the Sun at the centre point, is absolutely enormous.

Archimedes also notes that the new book was influential enough that Cleanthes, the head of the Greek Stoic movement, thought it worth advocating the prosecution of Aristarchus on charges of impiety. Galileo and the Pope were but the repeat show! Aristarchus was specifically accused of:

> 'putting in motion the hearth of the universe... [and] supposing the
> heaven to remain at rest and the Earth to revolve in an oblique circle,
> while it rotates, at the same time, about its own axis...'

The main difference to later events seems to be that Aristarchus was
never prosecuted, even if his 'book' seemed to disappear. Instead it was
the more muddled theory of the Pythagoreans in which both the Earth
and the Sun orbit another 'central fire' that was left to challenge the
everyday impression that we actually do live on a stationary rock while
the heavens and Sun whirl around us. But muddled theory or not, Plato
was very influenced by the Pythagoreans, and even hints (in the
Timaeus) at the idea that the Earth might be rotating on its axis. Two
thousand odd years ago, then, the idea that the Earth might move was
very much a live one.

That is until Aristotle explicitly refuted it. Paradigm shift! In a
particularly rambling account called *On the Heavens* he complains about
the Pythagorean claim that 'at the centre there is fire, whilst the earth,
which is seen as one of the stars, moves around it in a circle which
produces night and day...' What nonsense!, he says. Born from dogma,
he adds. The Pythagorean one that supposed that:

> '...the most honourable region belongs to what is most honourable, and
> fire is more honourable than the earth... and in consequence they do not
> think that earth lies at the centre of the sphere, but rather fire.' (*On the
> Heavens*, Book II, chapter 13)

Aristotle deals with the nonsense, not so much by challenging the role
of 'honour' in the arrangement of the universe, but rather by explaining
that 'what surrounds, that is "the limit", is more honourable than what
is limited.' That is Aristotle's method. He does not look for new evidence, rather he reinterprets the old. It is in a similar spirit that he
proves that the stars must be fixed immovably to their crystal sphere by
offering the new argument that otherwise we would hear them
moving. For most people, since Aristotle was the expert of experts, the
Father of both Science and Logic, and the authority on the Heavens,
these were logical truths requiring no further empirical investigations.

Indeed the history of science is the history of Aristotle. For hundreds of years the Church scholars considered him to be not only the
central philosophical figure—referred to not by name but more
impressively as 'the Philosopher'—but a source of revealed truth—his
status that of a kind of prophet. Islamic scholars still do. Key lessons
that Aristotle handed down the ages from this lofty perch ranged
across biology, theology, mechanics, and cosmology. His teachings
included significant factoids such as that:

* women do not have souls, nor even as many teeth as men;

* heavy objects fall faster than light ones, with the speed increasing in direct proportion to mass;
* the Earth is fixed motionless at the centre of the universe.

Opinions such as these were for centuries, due to his status, the indisputable 'scientific' fact. Or, as Karl Popper put it: 'the development of thought since Aristotle... remained arrested in a state of empty verbiage and barren scholasticism' ever after.

If Aristotle had any evidence for his views on female dentistry, it seems to have been lost. But at least as far as the Earth's position in the universe goes, he backed up his prejudices with plenty of arguments. These ranged from the fact that birds certainly seem to be able to fly normally in the sky, to miscellaneous weather observations. The killer argument for Aristotle, however, is the empirical one drawn directly from experiment: notably that a rock thrown vertically upward 'falls vertically downward, rather than slightly to one side, as it would if the earth was in any kind of motion'.

In fact, Aristotle reassured everyone, all objects, and not just rocks, behave in a very sensible way and make no movements other than that diligent effort to return to the centre of the universe. Since 'the Earth and the universe have the same centre', it follows inevitably that the Earth must be not only motionless in space, but also motionless on its axis. *QED*, with the rock.

And that's why, some four hundred years later when the astronomer Ptolemy constructed his cosmological picture, the Earth was placed securely at the centre, immovable as a mountain, just as it 'appears' to be in ordinary life. It's tempting to tut-tut at his credulity but, for many centuries, this Earth-centred model justified itself as a valuable tool for ships and navigators, and even for predicting celestial phenomena, such as eclipses. As the science writer James Gleick has put it:

> 'For so many good reasons the Earth was the centre of all things. Even when turning their gaze to the heavens, the constellations themselves turned round it in their patient, regular procession. Just a few bright specks in the firmament remained a puzzle—the planets—or the 'wanderers' as the Greek word implies—which indeed wandered erratically and freely amongst all the other stars.'

And then comes along a book. *De Revolutinibus Orbium Coelestrium* or, as it is usually translated, *On the Revolutions of the Heavenly Spheres*. Written by a modest Polish priest and part-time astronomer with an essentially theological desire to bring the heavenly 'wanderers' back into the fold. Copernicus did this by firmly placing all the heavenly bodies into circular orbits. In fact, it was only incidentally, as the price of this system tidying, that he set the Earth itself into motion, and made it just another planet orbiting an immobile Sun.

It's often forgotten that Copernicus was actually a Churchman. Yet it makes sense that he would be as, actually, whatever the popular perception, throughout the Dark Ages the Church had been an enthusiastic ombudsman for the new astronomy. That is why, far from seeing himself as a 'revolutionary', least of all a theological one, Copernicus saw himself to be an obedient and diligent Catholic (and one who took mathematical innovations, without attribution, from the Islamic astronomers) and dedicated his book to Pope Paul III, who received it all 'cordially'.

Cordially! But secretly, he sent an order to the Inquisition to arrest and torture Copernicus — right? That would make better cinema, certainly. But reality, at least in this case, was more prosaic. The 'Copernican Revolution' we hear so much about today, in terms of scientists overthrowing established orthodoxies, seems to have been conjured up, or perhaps merely misunderstood, by historians. As one philosopher of science, Bernard Cohen, puts it: literally, the revolution 'was not at all Copernican, but was at best Galilean and Keplerian'. But that does not have quite the same ring to it.

The Church authorities in Copernicus's time were quite prepared to countenance refinements of Ptolomy's (regularly updated) model of the universe. The drama came not so much with the Catholics but from the Protestant 'fundamentalists' who objected to all such mathematical explorations — and that battle for 'real Christianity' was yet still just brewing...

Anyway, back to the book. It opens with a special preface introducing the new theory. This bit was written by another Churchman called Augustus Osiander, who oversaw the printing of the first edition and who was in fact not a Catholic but a Protestant. In it, Osiander presented Copernicus's theory simply as a new 'mathematical' model for interpreting reality, rather than as an argument for changing views of the hierarchy of the heavens. (This was the sleight of hand that Galileo would always refuse to endorse.) Ironically, given history's verdict, far from being frightened of incurring the wrath of the Papal Inquisition, Osiander was fearful of opposing the leading Protestant theologians of the day who had warned publicly against any efforts to make the Earth revolve around the Sun. Martin Luther himself later condemned Copernicus for 'a fool' who ignored the Bible's description of how Joshua commanded the Sun to revolve around the Earth. So in such a climate, Osiander designed his preface to leave the questions of whether the heliocentric theory was a true description of 'reality' or not suitably obscured. And just to be on the safe side, he left it unsigned too! This, of course, left everyone assuming it was actually Copernicus who had written it, and that the evasive words were Copernicus's timorous attempt to downplay his 'revolution'.

Not that Copernicus was unconcerned about the book's reception. Indeed, he delayed the publication until after his death out of fear. But again, not out of fear of the Catholic Church, but in this case out of fear of the experts of the time — the academic authorities in the 'schools' who were, almost to a professor, devoutly 'Aristotelian' in outlook. Back then, professors were totally against free debate, as they still are, of course. After all, academics like to be experts, and they can't be that if the facts keep changing. Not to forget, courses might have to be rewritten!

Much better, instead, to have order reign in the centres of learning. And it did because, as the contemporary science writer George Sim Johnston has put it, 'Aristotle was the Master of Those Who Know', and perusal of his texts was regarded as almost superior to the study of nature itself. Crucial to Aristotle's method was to divide everything up into exclusive categories, and the universe comprised two worlds, the superlunary and the sublunary. The former was made up of the moon and everything beyond — and was by 'nature' perfect, imperishable, and unchanging. The 'sublunary' universe of humankind, on the other hand, consisting of the Earth including its atmosphere, was subject to change and decay.

And then along came Copernicus's theory which made a nonsense of this fine stuff. So stiff resistance was to be expected. Why give up a firmly established system for a new, unproved cosmology which contradicted common sense, as no less an empiricist than Francis Bacon said? You can't argue with that, so that's how it was left. Not so much with two worldviews at an impasse, but with the new one ignored and the old view left unchallenged. That is... until one Galileo Galilee appeared on the scene.

Galileo was not a great diplomat. In his acerbically witty *Dialogue on Two Great World Systems* (1632), he not only made it clear that he considered the defenders of Aristotle and Ptolemy, from the Pope to the Professors, to be intellectual clowns, but could not resist ridiculing fellow Copernicans too. He even had a go at his most influential supporter, Johannes Kepler, German mathematician, astronomer, and astrologer, because he had annoyed him by putting the planets into elliptical orbits,[1] and had speculated that the moon might exert a

---

[1] In 1609 Kepler produced his new model of the solar system in which the planets all travelled around the sun, with their rate of rotation varying as a consequence of their elliptical orbits. But no one remembers that!

mysterious 'pull' on the oceans. Now, tidal effects were Galileo's *speciality* and the Italian professor spared no punches. He has one of his imaginary characters in the book say:

> 'I am more astonished at Kepler than at any other... Though he has at his fingertips the motions attributed to the Earth, he has nevertheless lent his ear and his assent to the moon's dominion over the waters, to occult properties, and to such puerilities...'

And Kepler was on the same side! Galileo reserved his cruellest jibes for the process of making a clown out of Simplicio, one of the main characters in his dialogue. Simplicio's role was to be a silly mouthpiece for the new Pope, Urban V, and his views on the workings of the solar system. These were actually pragmatic — the Pope had various factions in the Church to placate. As late as April 1615, Cardinal Robert Bellarmine, as 'Master of Controversial Questions' at the Inquisition, wrote a suitably diplomatic letter which amounted to an unofficial statement of the Church's position. Which was that:

• it was perfectly acceptable to maintain Copernicanism as a working hypothesis; and
• if there were 'real proof' that the Earth circles around the Sun, 'then we should have to proceed with great circumspection in explaining passages of Scripture which appear to teach the contrary...'

Aided by this, Galileo breezed through his first trial under Pope Paul V, even flourishing his own new theory of the tides, which said that they are caused by the rotation of the Earth. Alas, even some of Galileo's firmest supporters could see that this was physically impossible. As to the main dispute, afterwards Galileo was merely warned not to 'hold or defend' the Copernican hypothesis, in the phrase of a 'certificate' Galileo obtained from Cardinal Bellarmine (but disputed by his enemies in the Church).

By contrast, the same Cardinal had arranged for another Copernican to be burnt at the stake in Rome some years earlier, a fact that is often used to add the darkest drama to Galileo's interrogation. Poor Giordano Bruno was tortured for years by the 'Holy Office' and delivered naked and bound to the Civil Authorities in Rome with a note stating that he was an incorrigible heretic, but that they should be merciful in their punishment. 'Merciful' though, in the Church's view, meant burning at the stake. This sad end to one of Europe's most original philosophical thinkers is often linked directly to the astronomical debate, but in truth it seems much more likely to have had something to do with the fact that Bruno was a rebel monk who had written many books, including one's in which he denied the divinity of

Jesus and argued that even the Devil would never face God's judgment. Those theoretical differences over the ordering of the solar system scarcely rated in comparison.

Galileo certainly knew all about this (as by an historical curiosity Bruno had once applied for the university post Galileo eventually obtained as professor of mathematics), but if he was alarmed he certainly hid it well. Instead, in 1632 along came a new, rather witty book lampooning Pope Urban.

It was publication of this that required Galileo, now an old, sick man, to appear a second time before the dreaded Inquisition in Rome. And this time the verdict was more empathic. At the close, Galileo was sentenced to 'abjure' the theory and to keep silent on the subject for the rest of his life, which he was obliged to serve out in... a pleasant country house near Florence.

As deterrents go, the sentence was scarcely terrifying. However, Galileo's books were put on the index of forbidden reading where they would remain for some centuries. It might seem all a ridiculous interference of politics and religion in science and it might seem easy to laugh at the people who put the Earth at the centre of the universe and made the Sun and the stars trot obediently around it—but in many ways this is the sensible way to proceed. After all, consider for a moment your own position in the universe. One minute ago—where were you exactly? In the same place? Perhaps if you are in a train you might say no, I was a twenty kilometres away! But if you are instead sitting quietly somewhere, it might seem ridiculous to insist that you were a hundred kilometres away, due to the Earth's supposed rotation around its axis. And stranger still to insist you were several thousand kilometres away, this due to the Earth's rotation around the Sun. Leaving aside the solar system's rotation around the centre of the galaxy and the galaxies headlong rush away from the original site of the 'Big Bang'!

Because, indeed, scientifically speaking, the real debate that was waged in the seventeenth century was over the nature of motion itself. To make sense of planets circling the Sun we have to accept Einstein's notion of them as falling through curved space-time, which is not in itself an immediately 'commonsensical' supposition. It's all about adopting a cosmic perspective.

Take the shift in our perception of the Earth created by the adventure of the Apollo moon landings. Here, it is not so much the 'fact' that we live on a speck in the universe as a new way to visualize the fact that counts. An image of Saturn and its rings taken by the Cassini probe captured by chance a far-away blue dot: the Earth.

However, it was a chance sighting of the crescent Earth 'rising' in the dark sky, glimpsed by the *Apollo 8* astronauts as they came from behind the moon after the first ever lunar orbit, that changed the way a whole generation felt about their planet. The poet Archibald MacLeish was moved to write:

> 'To see the Earth as it truly is, small and blue and beautiful in that eternal silence where it floats, is to see ourselves as riders on the Earth together...'

However, as far as changing perspectives goes, if today we often have in our minds a image of the noble Galileo gazing up at the night sky with his newfangled 'telescope' and discovering the four tiny moons of Jupiter orbiting their mother planet just as, he argued, the planets themselves must orbit the Sun, the reality was more ambivalent. Galileo's telescope was barely more powerful than a pair of sharp eyes on a clear night. At best, it maybe was the equivalent of a pair of binoculars today. At first no one else could see those moons, and if Galileo's telescope conclusively demonstrated the true workings of the universe, his contemporaries were entitled to ask why were his careful drawings of the moon all obviously wrong? 'The result of a fertile brain, rather than a careful eye', as one astronomer put it. But then, try to prove anything with a telescope made of spectacle lenses.

Figure 5. One of Galileo's sketches of the moon in 'Sidereus Nuncius', published in March 1610. The image obtained by using his new-fangled telescope contradicts not only the Ancients (who insisted that the moon's face was perfectly smooth) but that obtainable by the naked eye.

Galileo was an inveterate tinkerer, what today we might also call an 'early adopter' of all the new technologies. When he heard about the invention of 'spyglasses' in Holland he quickly built a simple telescope made of spectacle lenses mounted on a hollow tube, for himself, characteristically asserting full credit for himself for the contraption.

But though he invited other scholars to share his discoveries by peering through the spyglasses, few of them were able to see anything very much at all. Undeterred, Galileo used 'his invention' to shake the foundations of the Aristotelian universe. As Feyerabend says: 'Pointing it at the moon, Queen of the unchanging "superlunary world", he found it was pockmarked with craters and mountains and valleys — and not even a perfect sphere.' Where the Ancients, like Empedocles, had compared the moon to a large hailstone, made up not of frozen water but frozen air, Galileo noted that the object appeared to be 'rough and uneven' with mountains fully four miles high, and with deep valleys and chasms between.

The moon reduced to a rock, Galileo turned his spyglass to the heavenly fire that the Ancients had waxed lyrical about, and even Copernicus was proposing as the centre of the universe. Galileo reported that his spyglasses showed that the Sun's face was blemished — prone to sunspots! He notes this small 'paradigm shift' in a letter, saying: 'So long as men were in fact obliged to call the sun "most pure and most lucid", no shadows or impurities whatever had been perceived in it…' Now it seemed the Sun was in any case just one out of countless stars that no one had ever seen before 'in numbers ten times exceeding the old and familiar stars'.

Turning the spyglass towards Jupiter (the third brightest object in the night sky after the moon and Venus), he detected those four tiny satellites orbiting it and offered them as proof that no longer could it be said that all the heavenly bodies revolved exclusively around the Earth. And finally, and perhaps most damaging of all to the geocentric universe, he observed the phases of Venus, the only straightforward explanation of which was that Venus orbited the Sun and not the Earth.

The Church's response to these discoveries was mixed. On the one hand, the leading Jesuit astronomer of the day, Christopher Clavius, immediately acquired an improved telescope, for himself, and readily acknowledged both that Galileo was right about Jupiter's moons, and about the phases of Venus. The Vatican was not, however, prepared to accept all Galileo's conclusions yet, adopting instead as a kind of 'halfway' measure the cosmology of Tycho Brahe. In this system, all the planets except the Earth orbited the Sun!

This was a 'fix' worthy of the orthodox scientists themselves. But, indeed, it coped very satisfactorily with the new discoveries. So, when Galileo visited Rome, in 1611, the Cardinals were in good cheer and quite prepared to fete him as just another great Catholic explorer — in his case, of the night sky. That was why he was even granted a private audience by Pope Paul V, who offered his continuing support and good will. Paul explained that the Church was quite comfortable with

Copernicus's 'revolution' as an hypothesis, one perhaps even superior to the Ptolemaic system, but was waiting until further proof settled the matter before ruling in favour of it.

After all, one respectable argument remaining against putting the Sun at the centre of the universe, which had been discussed by the ancient Greeks as part of their debate over an Earth-centred universe and noted by Aristotle, along with all his own dodgy arguments, was that *if* the Earth did orbit the Sun, then small changes in the relative position of the stars should have been observable in the sky. That is, it should surely be possible to measure a shift in the position of a star when observed from the Earth as it orbited the Sun, starting at one extreme, and then six months later from the other.

'Parallax effect', as it is known, is not as grand or complicated as it sounds. You can experience the effect very easily by looking out of a window on one side of a room at, say, a tree, noting what the tree is immediately in front of, and then going to the next window, and seeing if the tree is still blocking the view of the same things. Which it won't be. This looking out when moving between different windows is the equivalent of observing distant stars when the Earth is at opposite ends of its 'hypothetical' orbit around the Sun. The distance between these 'windows', it was recognized, would be vast if Copernicus's theory was correct! Yet no matter how hard the astronomers looked, they could not see any change in the relationships of the stars — they looked like they were indeed all fixed securely to their crystal sphere. In fact, although Copernicus's theory was published in 1543, the first time anyone managed to measure any confirming parallax effect would be nearly 300 years later, in 1838.

Another significant, empirical weakness in Copernicus's theory was that it insisted (as did Galileo, even against the detailed mathematics of Kepler) that the planets orbit the Sun in perfect circles. The Jesuit astronomers could easily show that this was not in conformity with observations. But then Galileo was not really bothered about the details. What he was proposing through his observations was a funda-mental conceptual shift. No longer was there to be a single reference point — an unambiguous use for words like 'up' and 'down', but from now on he wanted there to be multiple reference 'frames', with familiar terms like being 'at rest' or 'moving' reduced to being only relative terms. Galileo was in fact the first of the Einsteinian relativists. For him, the moon was no longer floating above the Earth but was held there by a complex web of gravitational forces. For everyone else, it might have been easier to take him seriously if he had been claiming that the moon really was made of green cheese…

Actually, it was this esoteric lunar debate that Isaac Newton read about as a boy, in a book by a young chaplain, John Wilkin, called *The Discovery of a New World* (1638). It is in this debate, rather than the supposed experience of seeing an apple fall off a tree, that lies the inspiration of his theory of gravity, as well as the more precise realization that, over a certain distance, the embrace of gravity must become rather less powerful.

So how did this scholarly exchange descend to the level of the notorious 'Trial by Inquisition'? The explanation of that has more to do with human pride than with planetary motion. In reality, it was not Galileo's arguments that led to the chilling of his exceptionally friendly relations with the Vatican, but rather that 'Dialogue' he wrote lampooning the stupid person, called 'Simplicio'. Galileo's crime was more that of *laissé majesté* than *laissé Dieu*.

So where's all this history pointing? Just once again to the conclusion that science advances by irrational and political devices, not by weight of 'objective evidence'. And that even the Copernican Revolution was fought using propaganda rather than facts.

Even as a propaganda war, though, it's one that the 'scientists' emphatically won. In 1741, after a 'mere' century or so, rather grudgingly, the Church retracted its opposition to Galileo, when Benedict XIV approved the publication of the first edition of the complete works of Galileo. Just a shade less than two hundred years after Galileo's trial, in 1822, the Holy Office finally granted an official licence of approval to a work by Canon Settele, in which Copernicanism was presented as a physical fact and no longer as an hypothesis. For the Church, the battle for control of the stars was over. Yet the damage to the Church's intellectual credibility continued. Realizing this, after 150 years or so more reflection, in 1979 Pope John Paul II instructed the Pontifical Academy of Sciences to reopen the celebrated case, explaining that he felt the need to show to the world that for Catholics science has a legitimate freedom 'in its own sphere', and that, yes, this freedom had been unduly violated by Church authorities in the case of Galileo. And that is why, today, scientists are rarely burnt at the stake in Rome for their beliefs, or even summonsed for trial by the Vatican. A small victory for scepticism. But elsewhere, in the institutions where Copernicus's opponents really dwelt, scientific orthodoxy is more rigid than ever.

Chapter Seven

Black Holes, God Particles, and Bombast

Or why paradigm shifts are rarely forever

Subject: Quantum Mechanics (and other complicated sounding things)

What we're supposed to think:

'...this century has given rise to a "new breed" of physically trained philosophers in close contact with the technical side of physics and how it affects philosophical issues: like how to reconcile the tendency of macroscopic systems to approach equilibrium over time with the under-lying time-reversal invariance of physical laws; how to make sense of removing the infinities predicted by quantum field theory by "renorm-alizing"; and whether a plausible formulation of the "cosmic censor-ship" hypothesis holds true in general relativity so that determinism can be safeguarded against naked singularities.'

— Dr Robert Clifton, writing in the *Oxford Companion to Philosophy* (2005)

What you're not to say:

'Science alone of all the subjects contains within itself the lesson of the danger of belief in the infallibility of the greatest teachers in the pro-ceeding generation... Learn from science that you must doubt the experts. As a matter of fact, I can also define science another way: Science is the belief in the ignorance of experts.'

— Richard Feynman, physicist

Philosophers have a soft spot for quantum mechanics as (a) it sounds impressive and (b) within it normal rules supposedly do not apply. Out go boring things like Newton's laws of mechanics and 'Force = Mass *times* acceleration' and all that. All built upon commonsense assumptions such as that 'same cause must mean same effect', with the consequence that when the billiard ball hits the other billiard ball and causes it to move off it is with a velocity we can calculate — at least, that is, if it were not all so... boring! So thank goodness none of it is true. That is, at the quantum level.

And in fact quantum science's founding professor, Richard Feynman, agrees with sceptical philosophers like Paul Feyerabend that, although most people are very pleased with scientific explanations, scientists themselves (let alone philosophers) should not be. Feynman says:

> 'The more you see how strangely Nature behaves, the harder it is to make a model that explains how even the simplest phenomena actually work. So theoretical physics has given up on that.'

And although Professor Feynmann wrote a book called *Quantum Mechanics*, sure enough, that does not explain it either. After all, as he says in it: 'One does not, by knowing all the physical laws as we know them today, immediately obtain an understanding of anything much.'

In another tract called 'The Distinction of Past and Future, from the Character of Physical Law', he elaborates, saying that in all the laws of physics that we have found so far there does not seem to be any distinction between the past and the future. 'The moving picture should work the same going both ways, and the physicist who looks at it should not laugh.' In quantum science, it makes no sense to talk of past and future. But this sort of physics undermines our understanding of the world. Should physicists be doing stuff like that? Feynmann thinks yes, most certainly, as (like a latter-day Socrates) he believes:

> 'Science is the belief in the ignorance of experts.'

The true natural philosopher welcomes the fact that there are so many conceptual holes in quantum physics, which have great significance for, well, the study of black holes. Perhaps the biggest hole is that the theory of gravity as established by Einstein is incompatible with the habits of the quantum world. So quantum science simply uses a different version of how gravity works. Quantum mechanics and Einstein's theory of general relativity are the two great pillars of science but they are constructed to different specifications, and cannot be unified.

Another conceptual hole sucks in the mechanism determining the mass of fundamental particles. The problem? The masses of the electron, proton, and neutron are generated through what-is-called 'electro-weak breaking', but particle physicists do not know how this breaking mechanism works. Moreover, since all material is made of atoms and since atoms are made of electrons, protons, and neutrons, this unknown 'breaking' is given the job of deciding the mass of everything.

The most expensive scientific experiment in human history, the Large Hadron Collider, which squats underground somewhere near Geneva, Switzerland (together with its budget of 9 billion US dollars), is supposed to help answer this question. But, at least on that, we can be forgiven if we remain a little sceptical! Even the adoption of the speed of light to use as our universal benchmark, imposing order on space and time is problematic. Unfortunately, for something that is supposed to be the cornerstone of modern physics, not to say our understandings of the universe, the constancy of the speed of light remains debatable.

And then there is the 'horizon problem'. When astronomers point their telescopes at the cosmos, they find it looks pretty similar in which-ever direction they look—no matter how far away they peer. From one edge of the visible universe to the other, space is pretty much the same, not just in terms of stars and galaxies, but also more importantly (from the point of view of physics) in terms of the background microwave radiation that is thought to be the legacy of the Big Bang.

Radio telescopes suggest that the microwave energy filling the cosmos is at pretty much the same temperature pretty much every-where. Like a well-mixed café-au-lait, the energy of the original Big Bang has spread out evenly all round the cosmos. The universe has had an estimated 14 billion years to settle down, so that may not seem surprising—but, in fact, there is a rather big inconsistency that no one has yet been able to explain. Based on observation, the universe is thought to be at least 28 billion light years across, and since Einstein's theory of relativity says that nothing can travel faster than light, there has simply not been enough time for the laws of entropy to take effect and even out all the waves of microwave radiation—the hot and cold spots—created in the Big Bang.

In response, cosmologists have come up with some pretty desperate solutions. One is the theory of Inflation, which says that in addition to the Big Bang (which created the uneven background radiation), there must have been another effect almost immediately afterwards which made the universe suddenly blow up like a balloon. But no one knows what or who would have been responsible for doing the blowing. The knowledge gap is simply shifted along. Others have suggested that the speed of light used to be even higher, but has since slowed down. Again, no one knows why that might have happened. And another solution, of course, is that that there wasn't a Big Bang at all. Last but not least, there's the possibility that relativity is wrong.

Physicists don't like to talk about it, but as to that last possibility, even within their own circles there are plenty of doubters. After all, there is also the anomaly of the ultra-energetic cosmic rays. For more than a decade, physicists have been seeing cosmic rays that should not exist. Cosmic rays are little bits of atoms spat out by exploding stars somewhere outside our solar system that have travelled through the universe at close to the speed of light. According to Einstein's special theory of relativity, there is a maximum possible energy for these particles, but here's the disgraceful thing: over the past decade, researchers have regularly detected cosmic rays travelling well above said speed limit.

In fact, Einstein himself set out many problems with relativity in articles written towards the end of his life. In one, Einstein noted: 'If the speed of light is the least bit affected by the speed of the light source, then my whole theory of relativity and theory of gravity is false.' Similarly, Einstein warned: 'I consider it quite possible that physics cannot be based on the field concept, i.e., on continuous structures. In that case, nothing remains of my entire castle in the air, gravitation theory included, [and of] the rest of modern physics.'

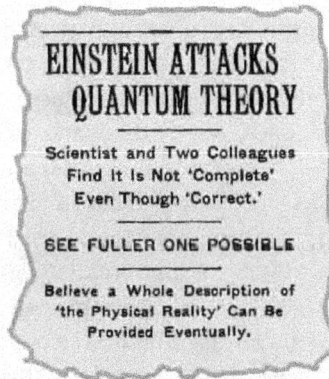

EINSTEIN ATTACKS QUANTUM THEORY

Scientist and Two Colleagues
Find It Is Not 'Complete'
Even Though 'Correct.'

SEE FULLER ONE POSSIBLE

Believe a Whole Description of
'the Physical Reality' Can Be
Provided Eventually.

Figure 6. 'Einstein Attacks Quantum Theory' read the *New York Times* headline of 4 May 1935.

We don't want that, surely. But Einstein's universe is in many ways a little bit of a castle in the air. If we think of it as made up of stars and planets and specks of dust swirling in a perfect vacuum, today's physicists think of it quite differently because only a tiny part of the universe is made of this sort of familiar material — the other 96% is still a mystery. Its presence has to be inferred instead from astronomical observations and the laws of gravity.

A Swiss astronomer, Fritz Zwicky, is credited with first noticing, in the 1930s, that the universe and galaxies seem to be held together by the gravitational attraction of a huge amount of unseen material — dark matter, dark energy, or dark fluid — call it what you will. In the 1970s more detailed astronomical observations by Vera Rubin of the size, shape, and spin of galaxies made the problem worse. That's another reason why today many billions of zlotys are being spent in order to hunt for dark-matter particles with the Large Hadron Collider at the European Organization for Nuclear Research in Geneva.

The Very Special Theory of Relativity

Unlike most major shifts in scientific thought, the Special Theory of Relativity was adopted by the scientific community remarkably quickly, consistent with Einstein's later comment that the laws of physics described by the Special Theory were 'ripe for discovery' in 1905. Max Planck's early advocacy of the Special Theory, along with the natural and elegant formulation given to it by Hermann Minkowski, contributed to its rapid acceptance among working scientists. Indeed, Einstein was much more open-minded about his work than his later supporters who instead preferred to 'stand on his shoulders' and decry all doubts.

Dark energy is today presented as an unexpected new discovery presenting a potential challenge to Einstein's fine theory which naturally did not allow for such mysterious entities. But the idea of an invisible unifying medium goes back many thousand of years before either Zwicky or Einstein. Every science has its 'demons', and this one has appeared in a multiplicity of forms from ancient times through the 'Enlightenment' and right through to the latest theories of 'quantum mechanics'. Electricity, gravity, atomic matter, and even spiritual life have called upon the services an invisible medium to fill certain theoretical gaps in their make-up. But, above all, it is light waves that have been better made sense of by supposing that, like water waves, they travel through something, in their case a 'luminiferous' something. Because 'dark matter', 'dark energy', is really what the Ancients called 'the aether'. But what is the aether?

To answer that, we need to go back to the beginning. To before the beginning, in fact. But here science draws a blank, and so imagination instead must take over. And for the Orphic priests at least, before the beginning there was Time. Not just any old time but a special kind of time called the Unaging Time, when nothing existed and nothing grew old. The Greeks called it Khronus — or sometimes Aeon to emphasize that it was an indeterminate and limitless time. And alongside the Unaging Time they placed Necessity, which they called Adrasteia.

Unaging Time and Necessity produced three new entities: one was the fathomless void which they called Chaos, another, which they called Erebus, was simply 'Darkness'. But the third, which they called Aether, was a blend of primordial Spirit and substance. And it was Aether that then copulated with her father, Time, to form and produce the Cosmic Egg. This was the first matter, solidified out of the infinity of the primordial mist.

Now the symbol of the egg appears in many creation myths, from Egypt to Persia to India. Or perhaps vice versa. But everywhere the earthly egg is the ideal symbol of new birth and new life. The special thing, from a philosophical, indeed a scientific, point of view, about this egg, however, was that it was also the first thing to be located in both space and time. Apart from that, it was reputed to be both gigantic and silver in colour. Naturally, when such a great, resplendent, silver egg hatched out, in a Big Crack, it was not the first chicken but the first god that sprang out. (And, as the broken eggshell split in two, one neo-Platonist philosopher adds, its parts formed Heaven and Earth.)

That's a fine story for children, you may say. Yet is the current physical model anything more? Anyway, rather than smash atoms in Geneva, let's instead examine some of the ancient myths for clues about fundamental reality.

When atoms are smashed, all sorts of strange things emerge. When the Cosmic Egg smashed, out came a Sun God with golden wings. It was also supposed to have four eyes, which permitted it to look (shine) in any direction, as well as to possess several heads in the shapes of various animals, the advantage of which is suitably obscure (other than that it enabled it to have the voice of both a bull and a lion). Although the Sun God was said to be invisible, it radiated pure light.

The Sun God was the first supreme ruler of the universe, but later, being an androgynous being and wanting a partner, it conceived and gave birth to Night (Nyx) to share the task. Some time later, the Sun God combined with Night, to create Earth and Heaven.

The story (as is the wont with myths) is rather complicated, but suffice to say that it was Night who eventually advised Zeus to swallow the Sun God. Zeus obliged and the entire universe including the other gods disappeared. With all this in his belly, Zeus gained new power and used this to create a new universe. A new Sun, new planets, stars, mountains, land, and seas were created. The other vanished gods were also reborn.

In the Orphic myths, the cannibalism of the gods is often apparent. Time swallows his children; Zeus swallows the Sun God and the entire universe; and other gods swallow each other. But the underlying idea is that of the cycle of existence — birth followed by death, which in turn is followed by rebirth.

That's the wisdom of the Ancient world. Or a fairytale, if you prefer. And most scientists do. But now then, what is *their* story? It turns out to be every bit as speculative.

Newton and the Green Lyon

Newton's laboratory notebooks are filled not only with details of how he deduced that white light is really a mixture of spectral colours, and sober explanations of optical and physical phenomena such as freezing and boiling, but with notes about green lions and dying toads, poisonous dragons and hermaphrodites. As William Newman says, 'whatever the ultimate purpose of Newton's alchemical investigations may have been, it is clear that we cannot erect a watertight dam separating them from his other scientific endeavours.'

John Maynard Keynes, who acquired the world's largest single collection of Newton's alchemical papers, went so far to say that 'Newton was not the first of the Age of Reason. He was the last of the magicians.'

Certainly, Newton would have not called himself a 'scientist', but rather a Natural Philosopher, a philosopher of nature. Such people were rarely if ever narrowly associated with a single discipline, as modern scientists invariably are; the leading physicists of the seventeenth and eighteenth centuries—the 'Enlightenment' period—were instead philosophers in the broadest sense; people who understood their theoretical and experimental work in the wider context of a worldview with religious and political elements included.

Take light waves, or radio waves if you prefer. To send these through a true vacuum would seem to require 'either' the existence of electric fields without associated electric charge, or electrical charge without associated matter... yet, as far as anyone knows, all electric fields require electrically charged matter. That is why scientists and philosophers originally invented the 'luminiferous aether', and that is why they have ever since been obliged to plunder many disciplines, including fluid mechanics, in search of concepts to explain how the aether could possibly work. Descartes headed off to fluid mechanics for swirling vortices and 'whirlpools'; Kelvin opted for the peculiar 'elastic-solid' properties of sponge.

It was the likelihood of light struggling to travel through either a series of whirlpools or worse still an 'elastic solid' that prompted one of physics' most celebrated experiments — the Michelson-Morley attempt to measure the speed of light. In this classic experiment, the two men accurately measured the speed of light travelling both 'with' the flow of the aether—and against it. And they found that, whichever way you measured it, light always seemed to travel at exactly the same speed.

Science has never been the same since. Yet, far from successfully 'disproving' the existence of the aether, as the experiment is today tidily said to have done, it rather aimed at supporting one version of the luminiferous aether over another one.

All this will seem like rather old wisdom, scarcely relevant to swapping experimental results at today's physics conferences. Yet there are many surprising similarities, between the 'new' physics and the Ancient variety, as Fritjof Capra has tried to communicate through popular books like *The Tao of Physics*. One key similarity is in the idea that matter comes out of energy, another is that the universe continually consumes and reproduces itself. (That's why Newton's notebooks also have sketches of copulating couples in them!)

Take that question of where does energy come from? Contemporary theories of the universe all start with the idea that the universe is a chemical reaction, in which stars powered from within. Otherwise, where would the energy that drives them come from?

This was the question asked by Sir Arthur Eddington in the 1920s, who settled for the tidier idea of stars as chemical reactions, and laid the foundation for subsequent mainstream models. However, the American engineer, Ralph Juergens, asked the question again in the 1970s, and opted for the latter complaining that the 'modern astrophysical concept that ascribes the sun's energy to thermonuclear reactions deep in the solar interior is contradicted by nearly every observable aspect of the sun'.

That there are problems with current theories was illustrated on the 20 January 2005 when the Sun produced a coronal mass ejection that achieved velocities greater than anything astronomers had seen before. It normally takes more than 24 hours for the charged particles of a solar outburst to reach the Earth, but these ones made the journey in just thirty minutes. Earth (some 96 million miles from the Sun) was immersed in what NASA scientists called 'the most intense proton storm in decades'. Proton storms get their name from the 'rain' of positively charged particles that can hit the Earth. A NASA headline article concluded, 'How they were accelerated, however, remains a mystery.' From a plasma perspective, of course, this phenomenon is less mysterious.

According to Juergens, stars shine because they are connected to electric circuitry within galaxies. An electric star's brightness thus depends on the power of the electric current feeding it, not on the amount of nuclear fuel available to burn.

Within this model, stars behave as anodes in a galactic glow discharge. And there are explanatory advantages to seeing things this way: many surface phenomena that can be seen on the Sun — hot corona, sunspots, prominences, flares, and so on — can all be explained by an electric Sun, but are more difficult to account for in terms of nuclear reactions. It is accepted that these do take place on the surface of stars, but not in the core, with the claim being made that this explains why the amount of neutrinos the Sun emits varies with sunspot cycles. The surface nuclear reactions are said to be by accelerating particles in an electric field — which is how experimenters produce them in laboratories.

However, if stars, galaxies, nebulae (and planets) are fundamentally affected by electric currents in the plasma through which they move, then it follows that they can change relatively quickly!

In the book *The Electric Universe* (Mikamar 2007) by Wallace Thornhill and David Talbott, there's a nice series of slides comparing ancient mythological symbols (from rock carvings) and then arguing that they all fit the structures observed in science as plasma electricity (see Figure 7).

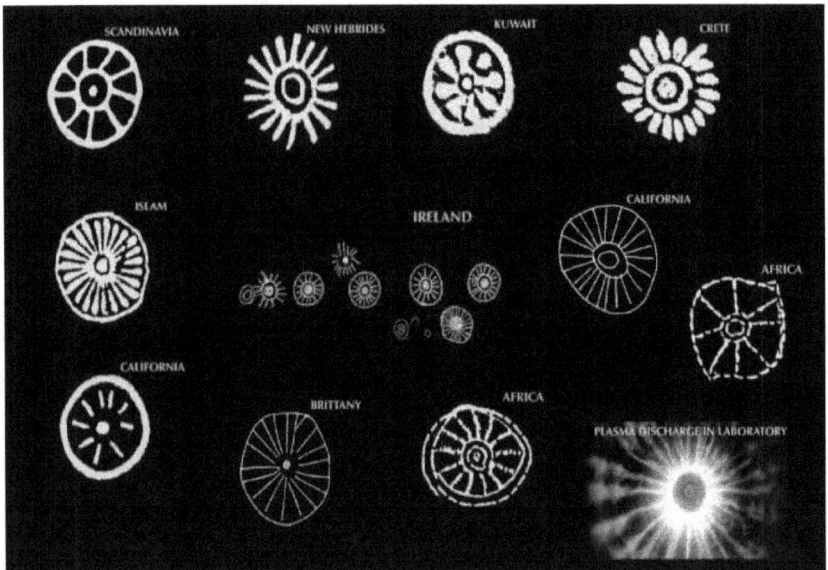

Figure 7. Ancient stone carvings and paintings illustrating supposed links between myths and the electric universe theory. (*Adapted from Thornhill & Talbott 2007.*)

Then again, there is a problem with binary stars. To begin with, why are there so many of them? In the plasma lab, of course, currents tend to run in braided pairs, but conventional astronomy has a trickier problem explaining this companionability. Sirius, the nearest and brightest star, has a partner, Sirius B, a tiny white-dwarf. The trouble is, when we look at Sirius B through Chandra, an X-ray telescope, it appears much brighter. How could this be?

Astronomers try to explain this in terms of gravity, claiming that particles fall into Sirius B so fast that the collisions emit X-rays. But charged particles don't care about gravity, so perhaps the better explanation is that the two stars create an electrical system in which the currents between them produce X-rays: a magnetic and a non-magnetic white-dwarf pair powered principally by electrical energy...

The grand idea of the original Big Bang is very much in the Platonic tradition of knowledge — that of starting out from idealized mathematical ('geometrical') principles. Such theoretical approaches, however, are fraught with problems, as the history of science testifies. After all, there is no *rational* reason to refuse that the universe has existed indefinitely, for an infinite time. It is modern myth that attempts to say how the universe came to be, either four thousand or *x billion* years ago.

In a collection of sceptical essays called *Science at the Cross-Roads* (1972), Herbert Dingle says:

> '...Lorentz, in order to justify his transformation equations, saw the necessity of postulating a physical effect of interaction between moving matter and æther, to give the mathematics meaning. Physics still had *de jure* authority over mathematics: it was Einstein, who had no qualms about abolishing the æther and still retaining light waves whose properties were expressed by formulae that were meaningless without it, who was the first to discard physics altogether and propose a wholly mathematical theory...'

Einstein introduced time as a physical dimension, and 'curved' empty space. Twentieth-century science has been obliged to seek confirmation of his theoretical innovations ever since. We should be doubly careful with such backward-looking attempts at verification. Take the famous tests to show time flowing at different rates made by the United States Naval Observatory in the 1970s, for example. These involved flying portable Ceasium-beam clocks in jet airlines. It made fine showbiz, but the test evidence is far from convincing, as the clocks varied so widely in time that some of the variations were more than the total supposed results during the time of the test. The most stable of the four clocks, no. 447, by itself constituting a better experiment than all the clocks together, indicated, as an overall result of the test, zero kinematic time accumulation.

Which brings us, or at least me and maybe Fritjof Capra, back to the often neglected subject of the ancient Hindu texts. According to these, the universe has two fundamental properties. These are motion and dimension—the latter being the space through which motion takes place. In Indian tradition this space is called the *Akasha*. The word, which appears in the *Rig Veda* in 1300 BCE, comes from Sanskrit and has the root *kash*, meaning to radiate or to shine. *Akasha* symbolizes a form of energy, a unifying energy which exists in every living being. And *Akasha* also has the meaning of 'aether' or the medium of movement. In both senses, it is indivisible, eternal, and all-pervading.

In Hindu mythology, *Akasha* is one of five elements that the human body is made of: the five being *Akash, Vayu, Jal, Agni,* and *Bhoomi*. Every feeling, both physical and mental, is due to *Akasha*. Indeed, all the secondary elements—solid, liquid, or gaseous—are formed by its unifying energy. In other ancient writings, such as the *Puranas, Akasha* is said to operate through sound or vibration. Thus it is identified with the element of Air, and in a spiritual sense with the 'breath of god', as it is put in the sacred Hebraic texts of the Jewish and Christian traditions. Where the book of Genesis says that the 'spirit of God moved upon the face of the waters', it recalls *Akasha* in action.

Is this science or mysticism? Don't be too quick to dismiss it as the latter. Consider the work of René Descartes, often (erroneously) credited as the first philosopher to apply logical rigour to his reasoning. Because it was Descartes who put forth the 'principle that interplanetary space must be a plenum occupied by matter imperceptible to the touch but capable of serving as the vehicle of force and light'. In other words, the aether.

Descartes considered medieval views on motion occult, and believed instead that all forces are transmitted by direct contact. To explain reactions between bodies not touching each other, such as two magnets, or the influence of the moon's position on the tides, he postulated that there must be some hidden, direct contact. Atoms, like billiard balls, required it. And invisible aether particles provided it.

In *Principia Philosophiae* (1644), as part of his rejection of the idea of 'action at a distance', the French innovator produced a new description of the universal based on everywhere being filled with an aether of tiny whirlpools exchanging even tinier particles with each other the whole time. The idea is that, owing to centrifugal force, each whirlpool is continually striving to expand, and so presses against the neighbouring whirlpools...

Descartes assumed that these 'aetherally whirling' whirlpool particles were in constant motion, but as there was to be no empty space for them to move to, they moved to places vacated by other aether particles. On the other hand, as we all know from those puzzles involving making pictures by moving squares in a grid, there must have been at least 'one' empty space somewhere... Or maybe there wasn't, in which case his theory (like some of those puzzles) simply didn't work.

Not to matter, in 1665 the publication of a new book, *Micrographia* by Robert Hooke, took Descartes' theory a little further by adding a new idea of light as a wave coursing through the aether, vibrating incredibly fast. And not long after this, Christiaan Huygens helpfully extended the wave theory of light by saying that the aether penetrates all matter and is even present in the vacuum.

In fact, Descartes had had a strong influence on the young Huygens, whom he had known when Huygens was a child. This helps explain why Huygens' aether was, like Descartes', made up of particles. Huygens' version specifically explained gravitation—a typical action without apparent direct contact—by saying it was due to aether particles rapidly rotating (in the manner of Descartes' whirlpools) in the space surrounding the Earth. Anyway, this then was the scientific landscape for Newton, as he started to develop his own theory, over in England at the time. No wonder, then, that Newton started his career as a strict adherent of aether theory. In 1672, writing:

> 'All space is permeated by an elastic medium or aether, which is capable of propagating vibrations. This aether pervades the pores of all material bodies and is the cause of their cohesion; its density varies from one body to another, being greatest in the free interplanetary spaces.'

Back then he had speculated that it might be friction with the aether that slowed a pendulum in a vacuum flask and even that it might be by drinking in the aether that the Sun was able to continue to burn. In 1675 Newton submitted a memorandum to the Royal Society in which, among other things, he too explained gravity in terms of the aether. He wrote that aether condenses continually in bodies such as the Earth, resulting in a constant downward stream of particles that 'impinges on gross bodies' and propels them through space. It is the movement of the aether, Newton suggested in 1675, that holds the planets in their closed orbits.

Of course, this represented but Newton's youthful foray into gravitational theory. Twelve years later, when writing the *Principia* (1687), he had become more inclined towards considering gravity as a mysterious example of action at a distance. He realized that this view would not go down well with many of his contemporaries, who had only just accepted that Aristotle might have been wrong to say that the reason an object falls downward is simply because it seeks its natural place in the universe. And indeed his apprehension was justified, both Huygens and Leibniz were very critical of his idea of an inexplicable gravitational attraction. So, when he came to write the second edition of the

Principia (1713), Newton defended his occult force by adding a 'General Scholium' in which he pointed out that his gravitational law was mathematically correct, even if he did not know the deeper reason for it. Now he briskly attacks the aether whirlpools of Descartes, and finishes grandly: *Hypotheses non fingo* or (less grandly) 'Hypotheses suck!'

Naturally, however, the question of the invisible unifying medium was not so easily dismissed. Problems concerning certain known optical effects — such as the interference of light or the simultaneous refraction and reflection of rays falling on the surface of water, meant that as the century came to an end it continued to haunt the new physics. Wave theory can account elegantly for such things, but theories, like Newton's, based on light as a stream of particles cannot. And then there were the questions raised about the speed of light on the cosmic scale, by the phenomenon of 'stellar aberration', discovered by James Bradley in the 1720s, after comparing the light coming from various stars and using the velocity of the Earth itself as a factor.

Sure enough, Huygens' approach was revived by Thomas Young and Augustin-Jean Fresnel to replace Newton's 'corpuscles' of light with waves once again ploughing through an aether. In addition (to explain the phenomenon of stellar aberration), Young decided that the aether had to be in a state of absolute rest, and finally (to explain how light waves can vibrate at right angles to the direction of travel, as observations showed them to do, but had been recognized since early in the nineteenth century to be possible only for waves travelling through a gas or a liquid), the two scientists made the invisible intangible aether into 'an elastic solid'.

As E.T. Whittaker put it in 1910 in his classic work on the history of the aether:

> 'The elastic-solid theory meets with one obvious difficulty at the outset. If the aether has the qualities of a solid, how is it that the planets in their orbital motions are able to journey through it at immense speeds without encountering any perceptible resistance? This objection was first satisfactorily answered by Sir George Gabriel Stokes, who remarked that such substances as pitch and shoemaker's wax, though so rigid as to be capable of elastic vibration, are yet sufficiently plastic to permit other bodies to pass slowly through them. The aether, he suggested, may have this combination of qualities in an extreme degree, behaving like an elastic solid for vibrations so rapid as those of light, but yielding like a fluid to the much slower progressive motions of the planets.' (*Theories of Aether and Electricity*, E.T. Whittaker, Longmans, Green and Co. 1910, p. 137)

By the end of the nineteenth century physicists had fine-tuned the theory of the aether as a quasi-rigid solid — not completely rigid because it can vibrate (Lord Kelvin's vortex-sponge aether) — luminiferous (light-carrying) medium that was completely massless, transparent, at absolute rest, and (last but by no means least) present everywhere. And it was now, by endorsing the ideas of Young and Fresnel, that the greatest physicist of electromagnetism of all time, James Clerk Maxwell himself, developed his Laws of Thermodynamics, laws that Richard Feynmann would describe as 'the most significant event in the 19th century.' Because, indisputably, Maxwell's equations are the unifying framework behind all of nineteenth-century electromagnetism, explaining the wondrous new technologies of electric light, electric motors, and even radio and X-rays.

They were also a way to represent a complex system of strains and vortex motions in the aether, understood as a tenuous but all-pervading medium through which light and every other electromagnetic disturbance was believed to propagate.

As ever with science, the theory was *almost finished*. There were just some details to be filled in. The necessity of the aether was self-evident, certainly to the British and Irish scholars. Only, the exact structure of the aether remained a subject for lively speculation. It was variously likened to a 'thin jelly' or a 'vortex-sponge', but not to any normal fluid because it had to be able to sustain transverse modes. Models were built and their merits debated at the regular meetings of the British Association or in the columns of *Nature* or the *Philosophical Magazine*. As well as Maxwell's own theoretical, 'imaginary' models there were intriguing 'real' ones, mechanical models made of wood and brass, with tiny wheels, pumps, and paddles.

One of the best was made by the eccentric Irish genius, George FitzGerald, whose uncle, George Johnstone Stoney, had the distinction of having named the electron. The electron was first conceived of by Stoney as a knotty point in the aether rather than as a charged particle. His uncle, like that other Irishman, not to say other George, the philosopher Berkeley, regarded the world of phenomena, and the motion of the elemental aether in particular, as a manifestation of the thought of God.

Be that as it may, FitzGerald had the universe as an array of brass wheels in a large array on a mahogany base with the wheels connected by rubber bands which were stretched and strained as the wheels turned. The movements of the cog-wheels represented the magnetic field, electromagnetic waves were modelled by oscillations of the wheels, while the behaviour of the elastic bands captured that of the electric field.

His model could be used to illustrate the discharge of a capacitor, and visualize the resulting flow of energy. Electrical displacement ran from regions of negative to positive strain in the elastic bands. Regions with no bands were perfect conductors. Mad? Foolish nonsense? Yet the wooden and brass models are no different from today's computer models, and their representation of the aether was in at least one sense accurate: exactly the same equations represented the energy of the model and in Maxwell's theory.

In addition, as one contemporary philosopher of science, J.M.D. Coey, puts it, to the nineteenth-century scientists the aether had a reality 'which we can hardly grasp' approaching, as it were, from outside the 'paradigm'. 'They believed that we are immersed in a medium in intense spinning motion, the equal counterpart of matter, and a manifestation of the omnipresence of God. No wonder those who had grown up with the aether and struggled to understand it were so loathe to abandon it!'

George FitzGerald was a formidable figure of the 'Renaissance Man' variety, whose eclectic interests spanned radioactivity, electrolysis, neurology, turbulence, sunspots, comets, and foam. He was actively involved in the design of electric motors for public transport, repairing X-ray generators, and projects to improve the quality of butter and rearing silkworms. Some years in advance of the Wright brothers, he had a prototype plane constructed in his College and undertook many, if rather short, test flights.

But out of it all, the beautiful model of the aether was his proudest achievement. Indeed, it was shipped to Nurnberg in 1892 at the expense of the Bavarian government to star in an exhibition of models and instruments used in pure and applied mathematics. Alas, its reign there was cut cruelly short. The work of Germany's own scientist, Albert Einstein, soon meant that the aether went out of fashion. FitzGerald's mahogany and brass model was taken off the shelves and put in a store cupboard. Around 1970 it briefly reappeared in public — tossed onto a skip outside the Physics Department.

The fate of FitzGerald's model signifies some sort of paradigm change... so to speak. Not least because, in 1881, it had occurred to the American, Albert Abraham Michelson, that it should be possible to observe effects of the aether on light waves. To do this, while visiting Germany, he had built a sensitive interferometer to determine the speed of the Earth with respect to the aether (in other words, to measure the speed of the 'aether wind') and compared the time it took for light to travel the same distance in two directions: either parallel or transverse to the Earth's motion relative to the aether. Despite his best efforts he could find no difference and was forced to conclude that the speed of the aether wind was zero.

The aether's believers at once rushed to its aid. H.A. Lorentz found an error and urged Michelson to repeat the experiment. This, Michelson, now back at home in Cleveland, Ohio, did, this time in collaboration with Edwin Williams Morley, a chemist also from Cleveland. But even using a new, better, interferometer they again found no effect from the aether wind on the speed of light.

Upon being informed of this second null result, H.A. Lorentz and George FitzGerald both independently postulated that rods fixed to Earth show a dynamic (i.e. due to a change in molecular forces) length contraction. This, they thought, was a better response to the new experimental evidence than abandoning the idea of the aether. After all, Heinrich Rudolf Hertz had supposedly just delivered 'empirical confirmation' of Clerk Maxwell's electromagnetic waves: it was universally agreed that the existence of aether was established. Definitely:

No paradigm change.

You see, the evidence for the old theory was just too strong. How could it be wrong? To be precise, Heinrich Hertz had demonstrated experimentally, between 1885 and 1888, the existence of the electromagnetic waves predicted theoretically twenty years earlier by Maxwell (finding waves from a centimetre to a metre in length, much longer than the wavelengths of visible light, which are measured in thousandths of millimetres). Indeed, Hertz himself stayed loyal to the aether throughout his life. His conviction of its importance grew, if anything, as a consequence of his research on electromagnetic waves, and eventually it became his chief preoccupation.

Hertz's experiments had an overwhelming impact on physics. They entirely overshadowed Michelson and Morley's aether-wind experiment. They raised the aether to the status of one of the basic building blocks of nature. In a letter to Hertz dated 14 August 1889, Oliver Heaviside, one of the founding fathers of electromagnetic theory, wrote:

> 'Then there is the vexed question of the motion of the aether. Does it move when "bodies" move through it, or does it remain at rest? We know that there is an aether: the question is therefore a legitimate physical question that must be answered. I am in hopes that extensions of your researches will supply material for an answer. As for the structure of the aether itself, that is a far more difficult matter, and one that can never, it seems to me, be answered otherwise than speculatively.'

Despite this, nowadays the Michelson-Morley experiment 'null result' is generally said to have disproved the existence of the aether. But the real story is both more complex and more interesting.

At the end of the nineteenth century the aether served two purposes. First and foremost it was the vital transport medium for electromagnetic vibrations. Secondly it offered an absolute frame of reference.
H.A. Lorentz loyally proposed that even if the aether underwent no
physical changes, all the properties required of it could be provided by
nothing more than its 'place' in space and time. This is a very philosophical response. However, within just a few years, one of the most
philosophical of all the great scientists, Einstein, was prepared to dismiss the aether entirely from consideration. His 1905 paper contains
just one reference to it saying:

> 'The introduction of a "luminiferous aether" will prove to be super-
> fluous inasmuch as the view here to be developed will not require an
> "absolutely stationary space" provided with special properties...' (A.
> Einstein, *On the Electrodynamics of Moving Bodies*, English translation by
> W. Perrett and G.B. Jeffery)

According to Einstein at this time (the period in which he was
developing the theory of special relativity), the need for a single universal frame of reference simply disappeared because the concept of
position in space or time was no longer absolute, but depended on the
observer's location and velocity. In another paper published in 1905,
Einstein made several observations on another, then-thorny, problem:
the photoelectric effect. In this paper he demonstrated that light can be
considered as particles that have a 'wave-like nature'. Since particles
did not need a medium to travel through, thus, neither did light.

This was truly the nadir of the aether, after 3,000 years of glory.
However, and very soon after, Einstein worried that his earlier dis-
missal of the aether might have been 'too radical'. And so, in a nimble
U-turn, he resurrected it for his theory of general relativity, writing:

> '...we may say that according to the General Theory of Relativity space
> is endowed with physical qualities; in this sense, therefore, there exists
> an aether. According to the General Theory of Relativity space without
> aether is unthinkable; for in such space there not only would be no
> propagation of light, but also no possibility of existence for standards of
> space and time (measuring-rods and clocks), nor therefore any space-
> time intervals in the physical sense.'

And in a paper called in German 'Grundgedanken und Methoden der
Relativitätstheorie in ihrer Entwicklung dargestellt' ('Fundamentals
and Methods of the Theory of Relativity'), published in 1920, he
admitted:

> 'Therefore I thought in 1905 that in physics one should not speak of the
> aether at all. This judgement was too radical though as we shall see with
> the next considerations about the General Theory of Relativity. It more-
> over remains, as before, permissible to assume a space-filling medium if

one can refer to electromagnetic fields (and thus also for sure matter) as the condition thereof.'

With general relativity also came the realization that 'empty space' could not be a giant electric field without changing, and so time had to be 'merged' into a new four-dimensional entity—space-time. This new space-time was, in a strange sort of way, every bit as rigid and absolute as Newton's abandoned 'Absolute Space', another foundational theory of modern science that Einstein is usually supposed to have done away with.

'Absolute Space' and the aether are both ideas that have been around for so long that their deaths have been announced several times, and yet it seems for both of them, even today, that such announcements have been premature.

The Pseudo-tensor

It was very early on pointed out to Einstein by a number of his contemporaries that his General Theory violated conservation of energy and momentum. So Einstein, in the manner of all great scientists, came up with an 'auxiliary hypothesis' to save his beloved theory. In this case, he invented his 'pseudo-tensor'. (A tensor is a mathematical expression used to manipulate geometrical entities like vectors. It has both magnitude and direction and is representable by a drawing of an arrow).

The sceptical Australian physicist Stephen Crothers, who has made a speciality of such things, argues that, first, since it is not strictly a tensor, it is not in keeping with the assumption in relativity theory that all equations be tensor in nature. Second, he objects that Einstein concocted his pseudo-tensor in such a way that it behaves like a tensor in only one particular situation, that in which he could contrive gravitational waves with speed c (that is, at the 'speed of light'). But the worst mathematical flaw is that the pseudo-tensor implies the existence of what pure mathematicians call a 'first-order intrinsic differential invariant'. The trouble with this is that such invariants do not exist! The mathematicians Gregorio Ricci-Curbastro (after whom the Ric = $R\mu\nu$ relationship that Einstein claims his field equations reduce to is named) and Tullio Levi-Civita proved that as early as 1900. So there appears to be an irresolvable contradiction at the heart of Einstein's mathematical case.

And finally, to those famous black holes which the latest telescopes offer pictures of—or at least reveal black areas of space. Einstein is often credited with predicting the phenomenon, yet he himself argued firmly against the possibility of black holes. The contrarian contemporary physicist Stephen Crothers argues that their existence depends (like Einstein's pseudo-tensor, *see box*) on a mathematical trick. In this case, on a derivation in which one of the complicating elements in the mathematics of Einstein's general relativity, the so-called energy-momentum tensor, is set to zero. Unfortunately, doing so essentially means removing all matter (mass) from the associated space-time. In essence, the mathematical theory from which the existence of black holes is derived describes an empty universe!

Physics has brought us so many incredible machines and insights that it is easy to award its practitioners god-like powers and abilities. But history shows that to do so is scarcely any better than worshipping the gods of the Sun and Moon. Maybe rather worse.

E Really Ought to Equal mc^2

The book *Einstein's Mistakes: The Human Failings of a Genius*, by Hans Ohanian, offers the following chronological list of Einsteinian 'slips':

1905 Mistake in clock synchronization procedure on which Einstein based special relativity

1905 Failure to consider Michelson-Morley experiment

1905 Mistake in transverse mass of high-speed particles

1905 Multiple mistakes in the mathematics and physics used in calculation of viscosity of liquids, from which Einstein deduced size of molecules

1905 Mistakes in the relationship between thermal radiation and quanta of light

1905 Mistake in the first proof of E = mc^2

1906 Mistakes in the second, third, and fourth proofs of E = mc^2

1907 Mistake in the synchronization procedure for accelerated clocks

1907 Mistakes in the Principle of Equivalence of gravitation and acceleration

1911 Mistake in the first calculation of the bending of light

1913 Mistake in the first attempt at a theory of general relativity

1914 Mistake in the fifth proof of E = mc^2

1915 Mistake in the Einstein-de Haas experiment

1915 Mistakes in several attempts at theories of general relativity

1916 Mistake in the interpretation of Mach's principle

1917 Mistake in the introduction of the cosmological constant (the 'biggest blunder')

1919 Mistakes in two attempts to modify general relativity

1925 Mistakes and more mistakes in the attempts to formulate a unified theory

1927 Mistakes in discussions with Bohr on quantum uncertainties

1933 Mistakes in interpretation of quantum mechanics (Does God play dice?)

1934 Mistake in the sixth proof of E = mc^2

1939 Mistake in the interpretation of the Schwarzschild singularity and gravitational collapse (the 'black hole')

1946 Mistake in the seventh proof of E = mc^2

Chapter Eight

Spooky Coincidences and Amazing Insights

Or keeping open minded about the implausible and inexplicable

Subject: Astrology

What we're supposed to think:

'Science has shown us through measurement, observation and experimentation that there are four forces in the Universe: electromagnetism, strong interaction, weak interaction and gravitation. For reasons too detailed to go into in this article, none of them can impact humanity purely from the positions of the stars in the sky or how aligned the planets are.'
— Dr Mark Thompson, astronomer

What you're not to say:

'In the early stages of the human mind, these connecting links between astrology and biology were studies from a very different point of view, but at least they were studied and not left out of sight, as is the common tendency in our own time, under the restricting influence of a nascent and incomplete positivism. Beneath the chimerical belief of the old philosophy in the physiological influence of the stars, there lay a strong, though confused recognition of the truth that the facts of life were in some way dependent on the solar system. Like all primitive inspirations of man's intelligence this feeling needed rectification by positive science but not destruction…'
— Auguste Comte, *Cours de Philosophie Positive,* (Vol. III, 1836)

One of the most surprising, some would say alarming, facts about Ronald Reagan is that, as soon as he became the President of the United States, he appointed a personal astrologer to help him take decisions. But then, for thousands of years, all the Kings and Queens had their personal astrologers to do much the same thing. These were the go-to experts on important state matters, such as when to invade the neighbouring country, when to harvest the crops — or how best to bring up baby.

Reagan had acquired the habit of consulting experts in the occult arts when he was but a humble actor in California, doubtless the process helped him decide which role in which film he should accept. We know where that led to, in due course: *Breakfast with Bonzo* (1951). Naturally, then, once he took high office, astrological advice could only become even more important.[1]

Early in his political career, Reagan scheduled his inauguration as Governor of California in January 1967 for just after midnight! It seemed the timing was dictated by celestial portents. Later, as President of the United States, Reagan routinely consulted a personal astrologer, Joan Quigley, about the personality and inclinations of other world leaders, and used these insights to help him assess the prospects of meetings succeeding. It seems, for example, that the stars looked favourably upon one Mikhail Gorbachev, the then leader of the otherwise Evil Empire, and hence Reagan was encouraged to attempt the rapprochement that in due course led to the end of the Cold War. In fact, the timings of all policy initiatives had to be squared with the movements of the cosmos, and White House staff were instructed to liaise with Ms Quigley in all their plans. She was responsible, in short, for the success of all that Reagan did. And these days, Reagan is counted as a pretty successful President, although that judgment is itself by no means necessarily a very scientific one.

Of course, Reagan came in for a bit of stick for consulting astrologers. Just as, more generally, scientists and attached pundits love nothing better that to mock more humble folk who follow their forecasts in the newspapers and magazines. But in this chapter, I hope to challenge certain socio-cultural prejudices that result from the triumph of a narrow sort of positivism.

Typical of the subject's many positivist detractors is one Mark Thompson, a writer and astronomy presenter for a TV show. Penning a little piece for a newspaper ('Astrology Rubbish? Don't Get Me

---

1   As revealed in Kitty Kelly's biography of the *Reagan Years*. Officially, Nancy had a lot to do with it, too.

Started', *Daily Mail*, 29 October 2010), he launches several torpedoes against the astrological dinghy.

First out, that claim that according to astrologers there are twelve signs of the zodiac. Wrong! Boom! Because there are thirteen signs of the zodiac, reports Thompson, before adding sardonically: 'Ophiuchus is the "new" one, yet for some curious reason I have never come across an Ophiuchian!' Ho, ho, ho… Similarly, he argues that until 1781 there were only five planets that could affect us, in the minds of astrologers: Mercury, Venus, Mars, Jupiter, and Saturn. Now there are three more! Oh no, two more, since in recent years the astronomers decided they were wrong to have ever counted Pluto as a planet as it is very small and appears to be really just another asteroid.

Even allowing for a moment the traditional twelve signs of the zodiac, there are problems. Mark was born in July which astrologers would say means that when he was born the Sun was in Cancer. But they're wrong again. Boom! Originally yes, the Sun *would* have been in Cancer when the star/Sun charts were produced thousands of years ago. But in reality, the wobble of the Earth on its axis — which astrono-mers call 'precession' — has led to them being all out of sync.

'In fact,' says Mark cheerily, 'when I was born, way back in July 1973, the Sun was in Gemini. *News Flash*: you're all reading the wrong star signs! All those astrology columns you've read that seemed spot-on were a fluke. Surprising eh?' But let's pause Dr Thompson there. Just think, what are the constellations? They are patterns in the stars that we see from Earth. The constellations are in reality made up of stars in very different positions in the universe. Typically, some of the stars are relatively close, and some are a long way away but may be rather larger and brighter. Some of the 'stars' may actually be distant galaxies! And the resemblance of the stars to anything like a lion let alone 'a goat with a fish's tail' is, well, nil. So where the planets are in relation to these imaginary, wholly human-constructed collections of bright points of light remains and may as well be wholly human con-structed. Anyway, it cannot make any difference to the claims of astrology where the 'physical' zodiac is. Because there is no physical sense in which the signs are really there in the skies.

Mark also offers more philosophical arguments, saying, for example, that if there is some mystical force (other than the funda-mental four known to physicists) affecting our lives from the planets, then clearly distance is no object for this force as it doesn't matter if a planet or star is near or far. And yet there are 'hundreds of exoplanets orbiting other stars?' And over 200 billion stars in the Milky Way! 'Surely that "force" would also be affecting us. Thankfully it doesn't, otherwise we would all be running round as complete loonies with all these "influences" flying at us from all directions.' Boom!

Mark could be said to illustrate the old saying about a little bit of knowledge being a dangerous thing. Indeed, he confides in closing, 'As you can tell, I'm not fond of astrology. It's all about telling people what they want to hear and we fragile humans' wanting something to believe in.' And that's bad, of course. And not just because it's *wrong*, either.

Mark is not alone. For many educated people, nothing better illustrates the gullibility and foolishness of the masses, and the need for the lead of a scientific elite, than the continued popular appeal of astrology. However, despite being resolutely logical and impartial, the expert critics don't seem to remember, or want to be told, that for a thousand years universities taught astrology as one of the core subjects, and that it was part of a sophisticated system of medical knowledge involving the different parts of the body and different herbs.

Nor do physicists really accept that the founding figure of their sensible science, Isaac Newton, only became interested in maths after he bought an astrology book at a fair in Cambridge, considered the subject one of the great studies of mankind, and that much of his writings are works that would nowadays be classified as occult studies. Similarly, astronomers don't want to think that their subject not only emerged but long profited from the mystical approach, far less do doctors want to hear that the best of modern medicine is borrowed from herbalism; or chemists that their subject is a side-shoot of alchemy. In short, in Paul Feyerabend's words, that everywhere science is enriched and sustained by unscientific methods and unscientific results. Instead, today astrology has firmly fallen out of favour with philosophers, let alone scientists. Nonetheless, thousands of years of thinking are contained in those ancient astrological myths and legends. Science is just a blip in this long history…

That said, one rather compelling argument heard against astrology is that some of its practitioners, people who compose astrology columns for daily newspapers or magazines, don't actually 'do it'. Instead of carefully checking the positions of the stars and planets each day, these astrologers simply write stuff off the top of their heads, or ask other people 'in the office' what they would LIKE to be their horoscope for that day, or week. Such glimpses behind the scenes look pretty damming, upsetting even for those who have up to then found a lot of thought-provoking insight into their most personal secrets in the relevant column.

Yet even here, the astrological beast that is slain is nothing more than the 'stalking horse' one of supposed physical cause and effect.

Although occasionally astrologers mutter about possible physical mechanisms for planetary influence, be they changes in the Earth's magnetic fields, or be they still more mystical 'quantum ones', really astrology is not about that. It is about mystical correspondences. It is no less likely that someone could randomly generate wise advice for an astrology column than they could do so using those out-of-date zodiacal charts.

Mystical correspondences are difficult to test in laboratories. Yet even if we allow that astrology's workings are too mystical to be examined, far less understood, by scientists, the same cannot be said of the question of whether their predictions are accurate or not. Here, scientists are on their home ground. It was very early on realized that, according to astrology, built upon the principle that our human characteristics are moulded by the influence of the Sun, moon, and planets at the time of our birth, twins should have had very similar aptitudes and characters.

There have been many more such scientific studies. Throughout the closing decades of the twentieth century, researchers tracked more than 2,000 people—selected as they were born within minutes of each other. Naturally, inevitably even, the scientists failed to find any evidence of similarities between the 'time twins'. They gleefully reported this in the *Journal of Consciousness Studies*, which is apparently where such studies belong, adding that: 'The test conditions could hardly have been more conducive to success... but the results are uniformly negative.'

Even the ancient Greek philosopher Carneades (219–129 BCE) had noted the issue, adding that if the crucial date for the influence of the stars on humans was the moment of conception, as opposed to birth (which, at least at the time he was writing, is necessarily always at least a few minutes apart), then the irrelevance of astrological influences for twins is even more marked. It was already clear back then that there was no answer to that, but equally, since Carneades was a philosopher who picked holes in everyone's claims to knowledge, and generally insisted that no certainty was possible about anything, in a sense he left astrology humbled but no more unreliable than any other study.

And sure enough, astrology quickly recovered from this blow to its prestige and soon was sufficiently influential that, a few hundred years later, the Church charged its top philosopher, Saint Augustine, with the task of attempting to discredit it. Augustine fell to with customary zeal and thoroughness. But his task was complicated as first he had to square frequent Biblical references to the stars (for example, guiding the wise men, who were evidently astrologers, to Bethlehem and the birthplace of the Messiah) with warnings elsewhere in the Bible that astrologers are evil people who belong in the category of sorcerers and witches.

It was truly a formidable exercise for any philosopher, and his solution was in effect to allow the trappings, but outlaw the substance. God did indeed use the heavens to send messages—but humans could not tell the future 'in general' by looking at the stars. At this point, Saint Augustine gratefully resurrected the twins argument as evidence that the claims of astrologers were false. And not only false. Worse! In trying to read the future, people sinned grievously. It was with this guidance in mind that Dante depicted the poor astrologers burning in the Inferno in a particularly painful way. He has them with their heads twisted viciously round, so that they can only see in the one direction — backwards.

Other Church Fathers (Jerome, Eusebius, Chrystostom, Lactantius, and Ambrose) all joined in the Crusade against astrology, and the great Council of Toledo prohibited it for all time. Many other critics in all ages followed Augustine's lead. By no means all of these were particularly devout, of course. Actually, astrology is one of the few things that religious folk and irreligious scientists can agree on. Roger Bacon (1214-1292), Marsilio Ficino (1433-1499), Giovanni Pico della Mirandola (1463-1494), William Fulke (1538-1589), John Chamber (1546-1604), and Sir Christopher Heydon (1561-1623) published critiques. By the end of the Renaissance in Europe, there was even a good market for parodies of astrology—even respected writers like Jonathan Swift and Benjamin Franklin had a go!

Of them all, Pico Della Mirandola is one of the most respected critics. And sure enough, his demolition of astrology rested upon a (re)discovery of the 'twins problem'. Actually, it seems that he had a particular grudge against the science, as he had once been advised by astrologers that he would die before the thiry-third year of his life, 'due to Mars and the direction of his ascendant'. Naturally he was pretty mad about that! And so he discredited their methodology. However, astrologers take unkind pleasure in noting that this unfortunate feature did indeed turn out to be fatal in his thirty-second year. That's the story, anyway, and indeed it's a fine story to frighten scientists with, but equally, as has been pointed out, even if the story is true, the prediction is still not really within the proper scope of astrological judgment.

So part of astrology's dodgy reputation relates to organized religion condemning it as a devilish art. For scientists, mine enemy's enemy is my friend. Yet even as Christianity condemned astrology with one hand, it embraced its traditions with the other. A trick it applied equally to all sorts of pagan and folk rituals and beliefs. That's why each Christmas the Church celebrates the 'sign' of the star over Bethlehem and has decorated its most important religious sites (in their masonry, stained glass windows, and paintings) with numerous astrological symbols and motifs. That's also why, six centuries on, the dates of popes' coronations were determined by the zodiac; aristocratic prelates employed their own personal astrologers; and signs of the zodiac appear all over church furnishings, tiles, doorways, manuscripts, and baptismal fonts. Such hypocrisy is worthy of science itself!

So there was plenty of scope for a new rediscovery of the 'twins problem' by science proper, in 1975, this time by a group of '186 leading scientists', including 19 Nobel laureates. In a preamble to their critique, the scientists explained their motivations. They had asked themselves why people believed in astrology and concluded that it must be that in uncertain times many long for the comfort of having guidance in making decisions. 'They would like to believe in a destiny predetermined by astral forces beyond their control. However, we must all face the world, and we must realise that our futures lie in ourselves, and not in the stars.'

The 186 were 'especially disturbed by the continued uncritical dissemination of astrological charts, forecasts, and horoscopes by the media and by otherwise reputable newspapers, magazines, and book publishers', saying it could only contribute to the 'growth of irrationalism and obscurantism'. The statement was organized by three experts in particular: Bart J. Bok, Professor of Astronomy at the University of Arizona; Lawrence E. Jerome, a 'Science Writer' from California; and Paul Kurtz, Professor of Philosophy at Buffalo, all regular 'skeptics' in the American sense of people who believe uncritically in the march of science and try to make money out of books discrediting all forms of esoteria and alternative medicine.

Lawrence Jerome, for instance, quickly followed up the petition with a 1977 book called *Astrology Disproved*, which offers as reasons why astrology cannot be true things like the good scientific fact that: 'Correspondence is not causation: no matter how powerful a coincidence, there is no necessary causation behind it.' But astrology is divination. Where tea leaves fall in the bottom of a cup does not 'physically' affect my future either. However, it is quite possible that there could be some significant correlations. Possible, but unlikely. One would need a tea-leaf expert to know. Astrology is from the same stable as the ancient Chinese art of throwing yarrow sticks and seeing how they fall—the *I Ching*. There is no possible way, gravitational or otherwise, that certain arrangements of yarrow sticks could influence events on Earth. Only the claim that certain patterns in the sticks reflect certain patterns in the universe. Patterns that ultimately we are part of. Yet listen to the positivist (everything real can be measured) mindset, plodding relentlessly though its irrelevant critique:

• Astrology is based on your time of birth, but if there is any planetary influence at all, shouldn't it be at your conception?
• There is more gravitational influence on a child at birth from the attending doctor than from all the stars and planets combined.
• Astrology is based on a Ptolemaic, Earth-centred universe and was left high and dry when the Copernican, Sun-centred model was adopted.

Nonetheless, these days, even if the world struggles to reconcile traditional religions with secular life, science has definitely become the final arbiter for *important* questions. And scientific method offers a way, no, more than that, *insists* on the way, to test the claims of its older sister, astrology. If astrology is right, then people should reflect their zodiacal origins. The trouble, of course, is that people are complex creatures, hard to categorize, and in any case, whatever their zodiacal inclinations, due to the complexities of life they may also have come to reflect many other things. Fortunately, however, philosophers set out their values and their characters in a uniquely clear way. They work for years to pin down for the benefit of others exactly what they think and why. Ironically (given their high-minded intentions), if any group should illustrate zodiacal influences, it should be these great thinkers.

A properly scientific account would surely include an effort to prove the hypothesis, as well as ones to falsify it. After all, discrediting things no one actually thinks or says is easy but worthless. So here is a little astrological experiment of my own, under the guidance of the excellent contemporary astrologer and thinker, Mark Shulgasser, to *scientifically* investigate whether or not astrology can tell us something — anything would do! — about humanity.

First, leaving aside supposed mechanisms, let us take as our experimental sample the special case of 'great philosophers', along with the evidence of their published views and writing, and see if there are any patterns or tendencies that astrologers might have anticipated — using only their crudest tool, the twelve zodiacal Sun-signs. Taking one hundred famous thinkers who I would call philosophers even if (owing in part to their own success) they may nowadays be called something else, such as 'scientist' or 'psychologist', there is a small but curious clumping of names under certain signs, and away from others. If you want to be a 'famous philosopher', it would seem best to be born under Aries (Descartes, Hobbes, Durkheim...) or Taurus (Kant, Kiekegaard, Hume, Mill...)! Philosophers with a starry-gaze bent should try to be born under Pisces (Copernicus, Galileo, Einstein...). Let me stress, though, this proves nothing. Proof is not what astrology is about. Instead, we should be content if we find interesting insights and make new connections.

In a standard, but otherwise unremarkable, astrological reference book entitled appropriately *The Zodiac*, C.E.O. Carter explains that the first three of signs of the zodiac – Aries, Taurus, and Gemini – have several peculiarities in common. He adds:

> 'There is something crude and undifferentiated about them. Aries repre-sents rude energy, such as we associate with the cave-man... Aries is the principle of Cosmic Individuality. This may also be called Cosmic Strength. It gives rise to the virtue of Courage, and as its counterpart it has the delusion of Egoism, or the belief that the rights of the individual are his only concern, the "struggle for existence" being deemed a perma-nent fact, beyond which it is not necessary to look.' (*The Zodiac*)

And indeed, starting this 'scientific' survey with Aries (22 March–21 April, ruled by Mars and the Sun), we immediately find a strange coincidence. Each sign has a motto, and the motto for Aries is 'I am'. Aries represents irresistible force, birth, emergence, sunrise. In terms of the body, it emphasizes the head. And Aries, it turns out, is also the sign of René Descartes, the great rationalist philosopher whose motto is... 'I think, therefore, I am': *Cogito ergo SUM, ego sum, ego existo.*

Descartes himself admits that the *cogito* is not a piece of reasoning. Paul Valery calls it 'a fist coming down on a table' and 'the explosion of an act, a shattering blow... If the *cogito* turns up so often in his work, if it is found again and again in the *Discourses*, the *Meditations*, the *Prin-ciples*, it is because it is an appeal to his essential egotism. He takes it up as the theme of the lucid Self; it is the clarion call to his pride and the resources of his being... I say that the real method of Descartes ought to be called egotism: the expansion of consciousness for the purposes of knowledge.'

The first thing astrologers would note about Descartes himself is, in Shulgasser's phrase, his rather splendid Mars quality. He carries a silver sword, he attends battlefields, he contributes to the mathematical physics of ballistics and munitions. On the Orleans road he disarms a rival lover. He has contempt for the past, the intellectual authority of the *scolia*, dead languages, even books. In his most famous work of philosophy, the *Discourse on Method*, Descartes uses the metaphor of battle: 'Or perhaps we should make the comparison with army chief-tains... for to try to conquer all the difficulties and errors which stand in our way when we try to reach the truth is really to engage in battle; and to reach a false conclusion on an important issue is to lose the battle.' To facilitate the goal 'to make ourselves masters and possessors of nature', he calls for this, perhaps his most tangible forecast of modernity, the organized campaign of science:

> 'Truth can be discovered only little by little... It is true that as far as the related experiments are concerned, one man is not enough to do them all; but he could not usefully employ other hands than his own, unless those of workers or other persons whom he could pay. Such people would do, in the hope of gain, which is a very effective motive, precisely what they were told.'

In other words, he calls for an army of mercenaries. He rejects volunteers, whose assistance would be 'at a net loss', for among other reasons 'they would infallibly expect to be paid... in compliments and useless conversation which would necessarily consume much of the time needed for investigation.'

There is a touch of testy impatience. Of, as Shulgasser puts it, the Generalissimo barking 'Do it now!' He feels pressure to see the practical results that will alleviate and transform Humanity. He's particularly concerned to expand human lifespan 'as much as a thousand years', if not 'prevented by the brevity of life'.

Blood flows around him. For a time, he purposely lives near a slaughterhouse. 'I have spent much time on dissection during the last eleven years, and I doubt whether there is a doctor who has made more detailed observations than I.' He is joined in this pursuit by several others born under Aries: thinkers delighted to pioneer the study of the mechanisms and vitality of the physical body. Sanctorius, born 35 years before Descartes, proposed that the body is a machine, and measures its temperature, rejects scholasticism. William Harvey, 18 years Descartes' senior, establishes the circulation of the blood. Descartes, disagreeing with Harvey about the heart's activity, cuts out part of the heart of a live dog and measures the pulsations along the aorta with his bare hand. He vigorously defends vivisection. He is credited with writing the first textbook of physiology but it follows on Sanctorius and Harvey, both also Aries; a team of them invade the body.

Descartes also frightens his neighbours and is thought to be an atheist. Pascal remarks:

> 'I cannot forgive Descartes; in all his philosophy, Descartes did his best to dispense with God. But Descartes could not avoid prodding God to set the world in motion with a snap of his lordly fingers; after that, he had no more use for God.'

Shulgasser says that it is a truism that Descartes introduces 'the Subject'. Aries invokes the immediacy of the nascent 'I', the human being-as-subject, striding frantically forth with a cry born of a state of emergency (and usually picking up the nearest object as a weapon).

'Astrology does not presuppose a simplicity of origin, but recognises in birth the agony and terror, as well as the miracle. In choosing a point representing the start on a circle or cycle, the necessary arbitrariness is a violence which provokes or uncovers a crisis-state. Descartes' *I AM* immediately confronts the Other: Doubt, the all-powerful Demon, and then opens a Pandora's box of dualisms, which are dealt with for centuries: mind/body, subject/object, self/other, conscious/unconscious, certainty/doubt. Not the least of them is progress/regress. The terror of regression, back into an imprisoning non-being, fuels Aries pre-emptive aggression with pre-rational force.' (Mark Shulgasser, 'Philosophical Rambles' in 2014, at www.philosophical-investigations.org)

Shulgasser says that his experience with astrology has led him to the position where he is sure it will continue to evade scientific proof so long as the rigid opposition of object and subject are fundamental to the concept of science.

'Astrology, it seems to me, is a tiny crack in the shell, through which the light of a nearly unimaginable new paradigm… glimmers, and it will remain for the foreseeable future as it is now, a curiosity and an amusement to the majority, while, for a handful, an obsession and bête noire.'

For the next 250 years after Descartes' birth, there was no need for another Aries in the philosophical sphere. There is, however, one other canonical philosopher born under Aries, and contemporary with Descartes, namely Thomas Hobbes, whose long life encompassed Descartes' birth and death. If Descartes originates the philosophical Subject, Hobbes does so for the political Subject. Hobbes is the man of two clear and distinct… not even sentences, clauses only: 'And the life of man, solitary, poor, nasty, brutish and short' and 'the warre of each against all'. Professor Gaskin sums him up well: 'menacingly terse.'

Think of Hobbes as Descartes' henchman. His synonym for 'anger' is 'sudden courage'; note the place of fear in this equation. Like Descartes, Hobbes is pursued for atheism, called 'the monster of Malmesbury' and accused by a Parliamentary Inquiry of causing the Great Fire of London. It seems only natural, for astrology, that Aries pugnacity would produce a unique style of hard-line atheism—now that there is no danger in it. And so to another interesting coincidence: all four of the four so-called Horsemen of the New Atheism, Richard Dawkins, Christopher Hitchens, Sam Harris, and Daniel Dennett, are Aries—not to forget also all notably belligerent. For Shulgasser, Dawkins with his selfish gene theory promotes a Hobbesian position, and as Oxford's Professor for Public Understanding of Science, he carries the sword of Descartes and the cudgel of Hobbes into the battle for the hegemony of science.

Indeed, it is a battle that would seem to scarcely need waging. Even in 1841, Charles Mackay had to track back several centuries in his popular, venerable, and rather long critique of such superstitious practices, *The Madness of Crowds*. Mackay's book was wildly successful precisely because it offered certainties in place of mysteries. One mystery debunked, for example, was the story of Nostradamus from what was surely the true heyday of futurology. Nostradamus, recall, was physician to King Henry II, husband of Catherine de Medici, born in 1503, at the town of St. Remi, in Provence, and one of the most famous astrologers of them all. His father was a notary and the son did not acquire much fame till he was past his fiftieth year. It was only then that his collection of verses, written in obscure and almost unintelligible language, began to excite the attention of the Royal Court.

Michel de Nostradame started life as a scientist, in the sense that he studied medicine. However, in those days medical training involved a hefty dose of astrological science, as each part of the body was understood as having a corresponding zodiacal influence. When France suffered terrible epidemics of plague, in the spirit of the great sceptics he stood apart from his fellow medical experts, who recommended 'bleeding' patients to remove the illness, instead emphasizing the importance of clean air and water. Wise though this advice was, it did not spare his wife and all his children from dying of plague in due course, and after this he travelled France collecting medicines and traditional herbal lore. Eventually though he returned to Provence, remarried, and settled down to write his famous obscure predictions. Some were based on traditional astrological calculations, but he also claimed to have direct 'visions' of the future. One of his early predictions foresaw the death of the King in a duel, and when this subsequently came about, his reputation was sealed.

Predicting deaths is a risky business, as current controversies about insurers having access to the analysis of individuals' genetic code echoes. Scarcely surprising if Nostradamus explained that he wanted his predictions to be obscure to avoid alarming people, as well as, doubtless, wanting to avoid incurring the wrath of the Church (which forbade direct prediction of the future).

The prophecies consist of upwards of a thousand stanzas, each of four lines. They are so vague, both as to time and space, that they are almost sure to be fulfilled somewhere or other in the course of a few centuries. If I predict that there will be a terrible earthquake, 'somewhere', in the next century, I am hardly going to be proved wrong. Like the best oracles, Nostradamus's code offers enough possible interpretations that the reader can find in it whatever they want to. The real work, in effect, goes on in the head of the seeker. Of course it is true that astrologers, like all the oracles promising to foretell the future, play strong on ambiguity and weak on specifics. That is both why their advice may not be right and yet they can still be useful to consult.

Such wishy-washy justifications are not enough for many professional astrologers and so they like to recall 'great predictions' of the past. Ones, that is, like those of perhaps the most celebrated astrologer of medieval England, a man called William Lilly. This man's esteem rested on the fact that he had successfully predicted the triumph of the Parliamentarians over the Royalists in the English Civil War, and later followed that up by foreseeing both the Plague of 1665 and the Great Fire of London a year later. Big predictions!

After this last event, a special committee of Parliament investigating the causes of the disaster summoned the astrologer to declare what he knew — and crucially, when. They were fascinated to see that in his book, *Monarchy or No Monarchy* — published in 1651, that is fifteen years ahead of the plague and fire — he had included a page showing on one side people in winding-sheets digging graves; and on the other side a large city in flames. A warning, evidently. And a more formal warning would have saved lives… why had the great astrologer not given one?

Naturally, in the manner of all oracles, instead of answering the question, Lilly replied merely by giving bland generalities, in this case a long speech in praise of himself and his astrological science. But when pressed, he also said that after the English revolution he had consulted his star charts to try to find out what might happen to the nation in general. What he read there he put into emblems and hieroglyphics, without any commentary, 'so that the true meaning might be concealed from the vulgar, and made manifest only to the wise'. 'Yes, yes, but did you foresee the year of the fire?', asked one literal-minded member of the committee. 'No,' replied Lilly, 'nor was I desirous. Of that I made no scrutiny.' After some further parley, it seems that 'the house found they could make nothing of the astrologer, and dismissed him with great civility.'

Make of that what you will. Another example, also related by Mackay, is, however, even more remarkable. It concerns the predictions of an Italian astrologer called Antiochus Tibertus, predictions that were often triumphantly cited by astrologers as proof of the truth of their science.

It seems that Antiochus uttered three remarkable prophecies; one relating to himself, another to a friend, and the third to his employer, Pandolfo di Malatesta. The first delivered was that relating to his friend Guido, at the time a much admired naval captain. It seems that Guido begged Antiochus to tell him his fortune, and so after much imploring the astrologer consulted the stars and the lines on his palm to satisfy him. It was with a sorrowful face that he then had to tell his friend that, according to all the rules of astrology and palmistry, Guido would one day be falsely suspected by his best friend, and should lose his life in consequence. Taken aback, Guido then asked the astrologer to tell him what in that case did the stars foretell of his own fate? So Antiochus again consulted the stars, and this time found that it was decreed that he should end his days on the scaffold!

Learning of all this cheery nonsense, Antiochus's employer demanded that the astrologer now predict his fate too, and to hide nothing from him, however unfavourable it might be. Of course, Antiochus had no choice but to comply, and sure enough, this time too, the stars produced an uncompromising message. It seemed that Malatesta, at that time one of the most flourishing and powerful princes of Italy, was soon to suffer great want, and die like a beggar in the commoners hospital of Bologna!

A Dogge miffing, where ?

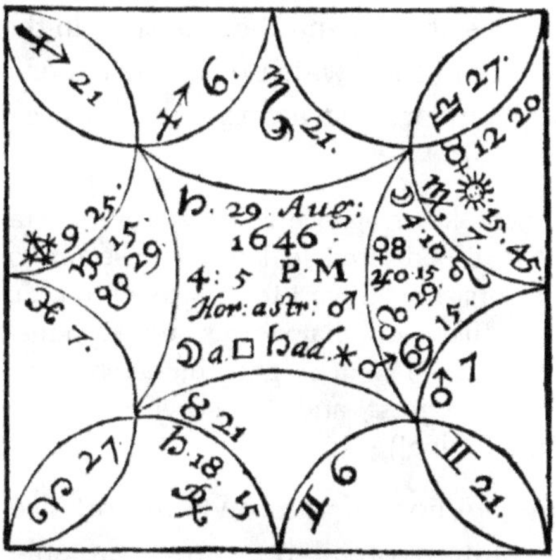

Figure 8. One of the celebrated astrologer William Lilly's charts. Lilly, called the 'English Merlin' by his admirers, was the last truly great astrologer of the West. This chart was drawn to find a lost dogge.

Sometimes horoscopes are best ignored. Yet, incredible though the predictions seemed at the time, in all three cases they became true. Not long afterwards, Guido was accused by his own father-in-law, the Count di Bentivoglio, of treason, and was assassinated as he sat at a friend's supper-table. Shortly after that, the astrologer was thrown into prison, suspected by his boss of having betrayed his friend. He attempted to escape, by letting himself down from his dungeon-window into a moat, but was discovered by the guards and swiftly executed on the following morning.

As for his employer, the celestial wheels were also silently working the prophecy's fulfilment. His city was attacked and seized by the Count de Valentinois and, in the confusion, Malatesta had barely time to escape from his palace in disguise. He was pursued from place to place by his enemies, abandoned by all his former friends, and, finally, by his own children. He at last fell ill of a languishing disease, at Bologna; and, nobody caring to afford him shelter, he was carried to the hospital, where he died.

This is astrology red in tooth and claw! This is what foretelling the future is all about! The only thing that detracts from the interest of this remarkable story is that it seems the prophecy was made up after the event. At least, that's what Mackay says. But then he would. Because Mackay sees hubris in our search for meaning in the stars, and in a sense he is right. But the reverse of his claim—that the universe, humans included, is but a soup of chemical reactions with 'consciousness' equally a myth—is a step that is harder to simply accept. Mackay has no time for theories of mankind that the stars in their courses watch over him, any more than they watch over worms or cockroaches, and 'typify, by their movements and aspects, the joys or the sorrows that await'! And he adds:

> 'How we should pity the arrogance of the worm that crawls at our feet, if we knew that it also desired to know the secrets of futurity, and imagined that meteors shot athwart the sky to warn it that a tom-tit was hovering near to gobble it up; that storms and earthquakes, the revolutions of empires, or the fall of mighty monarchs, only happened to predict its birth, its progress, and its decay!' (*The Madness of Crowds*, Chapter 6, 'Fortune Telling')

Indeed, an 'undue opinion of our own importance in the scale of creation is at the bottom of all our unwarrantable notions in this respect', says Mackay, teacher-like. So what is left after the pretensions of astrologers have been swept away? But there is something left. It is psychology. And appropriately it is Pluto, 'the messenger', that brings the insight.

The discovery of Pluto illustrates in a curious way both the role of the observer in affecting that which is observed, the issue which continues to vex quantum physics (where a particle's state is affected by who's looking), and the role of 'meaningful coincidence'.

Pluto is the ninth planet, or was until recently anyway. Recall that, throughout its golden age, astrologers worked with just five planets— and the moon—in their calculations. All of these were visible to the naked eye. In outward progress from the Sun, they are Mercury,

Venus, Mars, Jupiter, and Saturn. Earth itself nowadays slots in as number three, between Venus and Mars.

The first planet to be 'discovered' though was Uranus. It was found by William and Caroline Herschel (although Caroline does not normally get given any credit) on 13 March 1781, who noticed a 'star' that moved relative to all the others. Once spotted, Uranus was easily recognizable as a planet as it shows a disk when viewed through even fairly low powered telescopes. Astrologers were graciously prepared to give the new planet a role in their forecasts. Uranus was given the role of being the 'Awakener', the 'provocateur', the cosmic wake-up call for both individuals and society as a whole.

But it was soon realized that Uranus was not alone, however. The presence of another planet was hinted at by wobbles in the new planet's orbit, and sure enough, using ingenious mathematical arguments based on Newton's laws of gravitation, Neptune was found near where the calculations pointed. Again, astrologers had to revisit their charts and rewrite their guides.

Once the orbit of Neptune had been computed it was seen that the mathematicians had been quite lucky with their predictions. They identified Neptune's likely orbit correctly, but as it takes about 165 years for the planet to complete one orbit (it has not yet completed one since its discovery) it was only by chance that it was actually waiting in the section of the sky near Uranus ready to be discovered. First bit of 'luck' for the astronomers... Or was it more than that?

Anyway, all was still not satisfactory. For although Neptune is, by comparison to the Earth, a very large planet, what astronomers call a gaseous giant, it was still not enough to fully explain the irregularities in Uranus's orbit. Instead, careful study suggested that there must be another very large body even further from the Sun. Planet X, a hypothetical ninth planet, was assumed to account for the remaining wiggles. It was assumed to be another gas giant, but difficult to detect from the Earth because of the limits of the telescopes then available. For decades, the existence of Planet X was debated, but the question could never be resolved.

The interest led one wealthy and eccentric American mathematician, Percival Lowell, to devote a large part of his time, and his fortune, to the search. Working in a specially constructed observatory in Arizona, Lowell patiently checked a tiny area of space that mathematical calculations indicated to be where Planet X must be hiding. But Lowell died, after more than a decade of elaborate calculation and careful searching, with Planet X still undiscovered. Only a decade later, with better photographic equipment and telescopic instruments, would Clyde Tombaugh, working at Lowell's Arizona observatory, spot a tell-tale

moving speck amongst tens of thousands of other specks on photographic plates. The search was over, Planet X was located, and very
nearly where Lowell's painstaking calculations had predicted it would
be. Joke on the astrologers who again had to rewrite their guides.

But—and here's the rub—the new planet that Tombaugh found was
far, far too small to have exerted any kind of gravitational effect on
either Neptune or Uranus. Planet X, it turned out, was smaller even
than Earth's moon. As well as being of course a very long way away, it
is a very small rock and, even today, the most powerful telescopes can
pick it out as merely a dot of light. As one American astronomer put it:
'The question arises… why is there an actual planet moving in an orbit
which is so uncannily like the one predicted?… There seems no escape
from the conclusion that this is a matter of chance. That so close a set of
chance coincidences should occur is almost incredible, but the evidence… permits of no other conclusion…'

The mass of Pluto is now known accurately since a satellite, Charon,
has been discovered. Pluto's mass is 0.002 Earth masses, while Lowell
required Planet X to have seven Earth masses to produce the effects on
the other planets. That's what they call a significant difference. How
Pluto's eccentric orbit could have ever been accurately predicted is a
complete mystery, just like general questions about why the outer
planets fail to keep to their correct orbits.

After some weeks of confusion and puzzled rechecking of everything, it was decided that the discovery of tiny Pluto could be nothing
more than… coincidence. Astronomers decided to start looking for a
'tenth' planet instead. Decades later, with data from Voyager 2's 1989
fly-by, astronomers realized that the search for another planet was
unnecessary as the mass of Neptune had been inaccurately measured
all along and that the orbit of Uranus had never been unexplainably
perturbed after all. Coincidences like these are often dismissed as
trivial and insignificant. But the psychologist Carl Jung offers another
perspective on them.

> 'I know, from long experience of these things, that spontaneous,
> synchronistic phenomena draw the observer by hook or by crook, into
> what is happening and occasionally make him an accessory to the deed.'

Jung is remarkable and practically unique amongst the leading thinkers
of the twentieth century in that he explicitly acknowledges this type of
experience which countless human beings are very familiar with. His
descriptions allow the experiences themselves in their full integrity,
and give voice to them. After all, he had himself had powerful dreams
that he later considered to be prophetic. The alarming dream which
warns of the disaster or the premonition that saves a life do happen,

but they are fairly rare and the vast majority of coincidences are less dramatic and chiefly characterized by their apparent lack of meaning. Yet even so, because they are so bizarre, they take on a kind of significance and meaning.

Another kind of synchronicity is the meaningful coincidence between a vision now, a dream maybe, and events 'in the future'. Certain astrological predictions might well be considered to fall into this category. In the Foreword to a book on the ancient Chinese method of divination, the *I Ching,* Jung explains that synchronicity takes the coincidence of events in space and time as meaning something more than mere chance, namely, a peculiar interdependence of objective events among themselves as well as with the subjective (psychic) states of the observer or observers.

Jung starts from the ancient view that everything is linked, and that the world is the organic unity of all things. The alchemists' notion of the mystical *Unus Mundus* (One World) had a powerful influence on him. This view is also part of the most modern physics, if that makes it any more palatable to consider. Because everything is linked, the ancient alchemists believed that one phenomenon could mirror another — even if they were apparently quite distinct.

It's not that, for instance, the book opening at just the right page, the phone call from the friend you were thinking about at that very moment, matter in themselves especially, but that there is a 'meaningful coincidence'. This is the idea at the heart of the concept of synchronicity. And it is the notion nowadays frequently invoked to serve as an explanation of how astrology works. And indeed Jung himself did explore synchronicity in a famous 'marriage experiment', involving the statistical testing of astrological indicators of marriage.

Out of various possible astrological combinations for love or relationship, he selected three traditional factors from Ptolemy said to indicate marriage: Sun conjunct Moon, Moon conjunct Moon, and Ascendant conjunct Moon. He was so keen to analyse the results that he could not wait until all the data had been collected. Impatiently, he asked his co-worker to analyse the first batch of material which had been collected. The result appeared to be an extraordinary validation for astrology because the married couples showed a high contact of Sun conjunct Moon. Jung was aware that this would not be justified as statistically valid unless the rest of the material replicated the result. However, Marie-Louise Von Franz recalls that Jung was sitting in his garden musing over the results when suddenly he saw:

> '...a mischievous face laughing at him from the masonry of the wall... The thought struck him: Had Mercurius, the spirit of nature, played a trick on him?'

His instinct to distrust the result was confirmed by the ensuing joke with the remainder of the material. The second batch failed to confirm the result of the first one, but instead produced a high proportion of Moon conjunct Moon. The third batch came up with a high contact of Ascendant conjunct Moon and this finally washed away the significance of the first result. Taken overall, therefore, the combined results of the three batches showed no statistical significance for any single configuration. Yet for Jung, there was a significance here of a quite different order.

Jung failed to establish an objective correspondence between the heavens at birth and marriage partners. Instead, his own subjective state as an observer seemed to be involved, since the results curiously mimicked what he was looking for. As our scofflaw astronomer puts it:

> 'All those astrology columns you've read that seemed spot-on were a fluke. Surprising eh?'

Synchronicity postulates a meaning which is outside of and in some sense precedes human consciousness. Such an assumption is found above all in the philosophy of Plato, which takes for granted the existence of transcendental images or models of empirical things.

Since Jung's work on synchronicity theory, the 'new physics' has at least partially overturned the tidy world of science. The old Newtonian physics with its cause-and-effect determinism has been superseded by new perceptions of time, space, and the nature of existence. The new physics accepts that if we think we are dealing with objective matter, we will find to our surprise, as Jung did in his astrological marriage experiments, that we are not as separate as we had believed. That goes against commonsense, just as the idea that each of us might be mysteriously governed by the stars does too, of course. However, it is only in the realm of lifeless matter that scientists are prepared to swallow it.

Part III

Chapter Nine

Bubbles, Black Swans, and Banking Disasters

Or predicting stock markets with chaos theory. The danger of assuming an underlying regularity and order to the world, where in reality there is only uncertainty and unpredictability

Subject: Neuroeconomics

What we're supposed to think:

'The existence of analogies between central features of various theories implies the existence of a general theory which underlies the particular theories and unifies them with respect to those central features.'
—Josiah Willard Gibbs, unintelligible mathematician who influenced modern economics

What you're not to say:

'The ideas of economists and political philosophers, both when they are right and when they are wrong, are more powerful than is commonly understood. Indeed the world is ruled by little else. Practical men, who believe themselves to be quite exempt from any intellectual influences, are usually the slaves of some defunct economist. Madmen in authority, who hear voices in the air, are distilling their frenzy from some academic scribbler of a few years back.'
—John Maynard Keynes, *General Theory of Employment* (1936; 1947 edition)

In 2006, as the United States housing bubble entered its final throes of foolishness (with house builders giving away new cars to penniless future purchasers), the US Federal Reserve officials, tasked with monitoring the health of the nation's economy, merely saw danger from the threat of inflation—that the economy would grow too fast. Timothy Geithner, then in his role of President of the Federal Reserve Bank of New York, reassured colleagues when they gathered at the end of the year that 'We think the fundamentals of the expansion going forward still look good.' And there was earnest approval for Alan Greenspan, who had stepped down as chairman at the beginning of the year and by loosening credit controls had created the conditions for the boom that had continued throughout his tenure. Geithner even obsequiously gushed that his predecessor's greatness was still not fully appreciated, but predicted that it soon would be. A week can be a long time in economics. In any case, a year on the lush plaudits had become cruel brickbats. Writing on the causes of the US financial crash with the benefit of 20/20 rear-view vision, the OECD had no doubt where the blame lay:

> 'The policies affecting liquidity created a situation like a dam overfilled with flooding water. Interest rates at one per cent in the United States and zero per cent in Japan, China's fixed exchange rate, the accumulation of reserves in Sovereign Wealth Funds, all helped to fill the liquidity reservoir to overflowing. The overflow got the asset bubbles and excess leverage under way. But the faults in the dam—namely the regulatory system—started from about 2004 to direct the water more forcefully into some very specific areas: mortgage securitisation and off-balance sheet activity. The pressure became so great that that the dam finally broke, and the damage has already been enormous.'

Forever dear to the heart of many economists is the idea that, even as it is built up from millions of individual and possibly irrational decisions, there are certain scientific laws that govern economic behaviour. Discovery of these laws is the real Philosophers Stone—it really would turn base metal to gold. Economists offering politicians such laws, or analysts offering investors supposed advance predictions, are modern-day alchemists specializing in the darkest of dark arts.

Yet the primary conclusion of economics' central theory, appropriately called 'the General Theory', is that national economies are inherently unstable, largely because of the irrational expectations which drive them to exuberance sometimes—and into deep depression at others. Inherent instability and unavoidable contradictions mean that although, in a sense, the science of economics is based on mathematics, serious economists can legitimately give completely different answers to the same questions. That's why they're not sure whether to put interest rates up, or down, to protect or devalue the currency, to pump money into the economy or push up taxes and damp down demand. All over the world, as Keynes says, practical politicians, from rich industrialized nations to poor nations dependent on subsistence agriculture, ask abstract theorists to help them to box shadows.

Masters of the Universe

Emanuel Derman, professor at Columbia University, says that economics is the study of how to utilize limited resources to achieve good ends, a quintessentially rational activity. And good, he says, is in the eye of the beholder, defined by humans. Well, that's a philosophical debate. But in any case, as he points out, economists don't agree with each other about ends or means. They don't agree on the efficacy of printing more money to boost economies in hard times—or tightening public finances and imposing austerity. As he told one newspaper:

'They keep changing their minds every few years about conventional wisdom while at every instant appearing to be certain that they are right. My gripe with economists is not that their models don't work well—they don't, look at the role of central banks in the financial crisis—but that they seem so reluctant to acknowledge the riskiness of their advice. And yet, beware their fearsome unelected power.'

But if you really want to see economics in action, take a look at the financial markets. It is here that the worth of the world is weighted and judged. It is here that private and public fortunes are really made or broken. And it is here, amongst the shimmering screens of digits and ever-changing charts, that economists are indisputably king.

At the time of the last stock market crash, rich investors were enthusiastically employing ever more sophisticated computer experts, and their computers, to invest their money for them. The computers, programmed with esoteric principles and rules, whirred through millions of possible permutations of events, epitomizing the triumph of rational economics. The fact that only the most nerdish mathematicians could make any sense of what was going on inside them only added to the sense of progress.

Year after year new Quant funds (short for Quantitative Investment Funds) generated huge profits. These were really the smart place to put your money. By 2007, the combined assets of the Quants specializing in US stocks were comfortably over $1.2 trillion. Trillion mark you, not billion. That's a hundred dollars for every man woman and child on Earth. Or put another way, if investors all had a thousand dollars to put into a new fund, it would need one thousand million members. Imagine the annual shareholders' meeting!

And then came the crash. In 2008, one of the United States' biggest investment banks, Bear Stearns, 'blew up', as economists like to put it, and suddenly all the computer models failed to cope. In not much more than a year, two thirds of the $1.2 trillion had disappeared. Money mangers of Quant funds are still not quite sure where it all went. Some accuse their clients of 'panicking' and forcing the funds to sell investments at a heavy loss, 'evaporating' the wealth. But others say that it is the computer models themselves that had a fatal flaw. Because they expected the future to resemble the past.

If so, that's an assumption we all make. We insist the Sun will rise tomorrow just because... it always has done. We know it's dangerous to jump off window ledges on Madison Avenue because the laws of gravity apply just as much today as they did yesterday. Yet, in fact, there's nothing logical about such assumptions. In philosophy it's called the fallacy of inductive reasoning. But equally, anyone who doesn't regularly employ the fallacy would seem to be pretty irrational.

Certainly, when it comes to investing, most people intend to act rationally, in the sense that they want their actions to lead to a particular end — they want to end up with more money than they started with. That's why the smart investor prefers 'familiar' stocks to 'unfamiliar' ones; prefers to buy what everyone else is buying, and looks for 'patterns' in the movements of share prices. End result? Most investors buy high-street banks and big name brands, and prefer their own country's companies. They buy stocks at the peaks (noticing something has become popular) and are totally unprepared for 'the unexpected'. Does that matter? It does if you didn't actually intend to go bust. Because stock markets, like economics, do not respect our common-sense notions. The rational investment, just as much the rational investor, is a mythical creature.

It turns out that when it comes to making wise investments, acting either as individuals or as governments, basically... we don't. We follow the latest fad, we follow the herd, we act essentially irrationally. As newspaper stock market commentator Joe Nocera has observed, individual investors spend hours on internet chat boards (where the herd mentality is fiercest), adhere emotionally to sinking investments rather than cutting their losses, and then gamble their shirts on hunches and high-risk personal favourites. Ah, why do we do that when it's so foolish? But it seems that we human beings are in a sense 'hard-wired' to be irrational. It's a side effect, as the psychologist Clotaire Rapaille has argued, of long ago being mere reptiles, with brains evolved simply to seek out food and avoid known predators.

None of this is particularly new or controversial, even if over the years it seems to keep having to be 'rediscovered'. In the 1990s, it was even all the rage — under a glossy new name: behavioural economics. With the dawn (and new stock market crashes) of the twenty-first century, it became 'neuroeconomics', a study supposedly bringing together all the expertise of three sciences — biology, psychology, and economics. But whatever it is called, the central finding always remains the same. Humans have evolved, successfully one would suppose, using certain survival strategies (such as not going in dark holes and making sure to hang a little bit back from the front when travelling in a group). More subtly, we have learnt to identify patterns and regularities in the world, like the fact that certain berries make us sick and certain berries don't. But, as the 'behavioural economist' Zweig puts it, 'when it comes to investing, our incorrigible search for patterns leads us to assume that order exists where if often doesn't.'

That's where *neuroeconomics* comes in. Of course, it's not just economists who construct complex models both to explain and to help predict future events. In social science, for example, theoretical and mathematical models try to explain the way societies work. A model here is a mathematical construct which, with the addition of human interpretations, describes observed phenomena. The sole justification for such 'mathematical constructs' is precisely that they are expected to work, as the twentieth-century mathematician John von Neumann once put it. Yet, even when the models ingeniously do match reality (as tested against past and present situations), they remain at heart just abstract theories that seemed to work last week. And sometimes the world changes.

It is economics — and those stock markets — that illustrates the ancient philosophical problem of change best. Conventional economists assume that, whilst there is a buzz of random short-term changes in stocks and shares, the long-term trends are determined by sensible, macro-economic (that is, large-scale) factors such as changes in technology or productivity, new inventions, or wars. They assume that prices change smoothly, rather than in abrupt jumps — an assumption borrowed from the physics of movement. And indeed, most of the time, economics and the markets themselves do behave. But not always. Suddenly, and without warning, they can crash.

That's what happened on 'Black Monday', 19 October 1987, Wall Street's worst ever day, at least, that is, 'at the time of writing', and again in March 2000 when the internet bubble burst — and again during 'Black Week' (literary flair and reporting economics are not natural partners), September 2008, when the not-particularly-hard-to-spot problem with doomsday policies of easy loans to poor people for expensive houses came home to roost. Mighty banks crumbled and stocks crashed almost overnight, and millions of people woke up to find that they'd lost lots of their savings.

And none of the 'experts' had had the decency to warn them in time to do anything! Or that's how it seemed, because the nearest thing to an inviolable rule in stocks is that it is the small investor who gets burnt while the big fish somehow manage to get wind of events, and even turn them to their advantage. Sometimes it seems almost magical — how did Goldman Sachs know to disinvest in BP just before the oil company's drilling rig exploded in the Gulf of Mexico, not only causing America's biggest ever oil slick and not only killing eleven workers, but horror!, causing the oil company's share price to plummet? And sometimes it seems less magical than plain suspicious — how did long-term institutional investors supporting some company come to sell their investments just a week or two before the crisis announcement that dramatically affected the company share price...

Who Will Regulate the Regulators?

One of the most ancient themes of political and moral life, *quis custodiet ipsos custodes*... who will watch the watchers. Or in this case, the Wall Street regulators.

After the 2008 financial crisis, the New York Fed, now the chief US bank regulator, commissioned a study of itself. The study to understand why it hadn't spotted the dangerous and destructive culture dominating life inside the big financial institutions, let alone step in to intervene before it got out of control. The 'discussion draft' of the Fed's internal study, led by David Beim, a Columbia Business School professor and former banker himself, was sent to the Fed on 18 August 2009.

Given that the crisis cost US taxpayers billions if not trillions of dollars and wrecked many lives, you should not be surprised to learn that the report was intended to be kept entirely secret from the public. However, the contents were leaked to the press, and the gist was this: regulation of the banks failed because the banks' employees were discouraged from asking questions, speaking their minds, or even just pointing out problems.

On the contrary: the Fed encouraged its employees to keep their heads down, to obey their managers and, in their dealings with the big banks and investment houses, to not ruffle any feathers. The report quotes Fed employees saying probing things like, 'until I know what my boss thinks I don't want to tell you', and 'no one feels individually accountable for financial crisis mistakes because management is through consensus.'

Yet when the whole stock market crashes, everyone, even the giant transnational insurance companies, even the most select 'hedge funds', gets burnt. Even those so-called 'Quant funds', for elite investors, get caught out.

So what did the economists and other experts have to say following market crashes? Searching for reasons for 'Black Monday', analysts could find only 'herd behaviour' to blame. They could hardly come up with any other explanation, as the very next day the herd was back at its consoles—only this time buying stocks and shares with record breaking zeal. Should we say that the canny strategy for investors is to be irrational animals after all? Well maybe we should. But that's a difficult pill for economics to swallow.

After all, economics is a tidy universe at heart. The word itself comes from the Greek *Oikonomia* which is a combination of *Oikos* 'house' and *nemein* 'manage'. In standard economic theory, everyone will hustle down to their nearest budget store to stock up on unbranded goods, tirelessly computing the savings on all. It's not rational to want 'branded' beans or shirts although, maybe, it's 'aesthetic'. Purchasing decisions may be influenced by ethics, by social values, by politics. Worse still, collectively, the decisions themselves influence social values, politics, ethics. 'Everyone's got to have the new smartphone.'

But then economists, as the saying goes, are people who know the price of everything and the value of nothing. For most of them, why you buy a tin of beans is not important, it only matters that you do so. Yet in real life, making choices is the crucial stage. Think about your own purchasing decisions. If you have a choice between two pieces of cheese, one that is at 'normal price' and one that would be more than double the price of the other but is marked as a special buy at half price —which one would you buy? I know which one I'd go for. The expensive one under the offer. I'd assume it is better. I'd not even consider that maybe it's a rotten cheese which needs a lot of marketing... People do similar things with shirts, or haircuts, or college courses.

The study of economics bears the imprint of that canny Scottish philosopher Adam Smith, with his theory that human society is governed by that great implacable force, called 'self-interest', and that it is this force that shapes economic theory. Smith explains all in his barnstormer of a book, *The Wealth of Nations* (written in the economic dark ages in 1766, but still a popular read with right-wing politicians), where he offers that people are just cogs in a beautifully functioning money-making machine. Famously, it is 'not out of the benevolence of the butcher or the baker' that we can expect our supper, but only from their enlightened notion of their own self-interest. Self-interest arranges people into the splendid wealth-generating machine that is modern society.

'The wheels of the watch are all admirably adjusted to the end for which it was made, the pointing of the hour. All their various motions conspire in the nicest manner to produce this effect. If they were endowed with a desire and intention to produce it, they could not do it better. Yet we never ascribe any such intention or desire to them, but to the watch-maker, and we know that they are put into motion by a spring, which intends the effect it produces as little as they do.'

However, Smith is well aware that economics is not *just* about money. In another, less-celebrated, book he writes:

'...it is chiefly from this regard to the sentiments of mankind that we pursue riches and avoid poverty. For to what purpose is all the toil and bustle of the world? What is the end of avarice and ambition, of the pursuit of wealth, of power, or pre-eminence?... To be observed, to be attended to, to be taken notice of with sympathy, complacency, and approbation, are all the advantages which we can propose to derive from it. It is the vanity, not the ease or the pleasure, which interests us.' (*The Moral Sentiments*, I, iii, 2, 1)

Smith is considered in many ways to be the 'father' of modern eco-nomics, yet personally he put human values ahead of monetary ones. Economists ever since have done the opposite. If Liberal political philo-sophers such as Smith but also, a generation later, J.S. Mill, felt it almost intrusive to look at what individual consumers did or did not do, they nonetheless all agreed on the duty of politics — to 'enable' everyone to exercise as much free choice as possible — typically by spending money and making purchasing decisions. That's why, right up to the early part of the twentieth century, economists spent considerable effort trying to understand household needs and expenditures, and looked at con-sumption expenditure trends by income class. But 'fashions change', and by the middle of the century no one wanted to study family budgets. Instead, everyone wanted to discuss grand theories of how economic forces drive states and money itself. Two of the grandest were the economic theories known as 'general equilibrium theory' and 'Keynesian macroeconomics'. Neither of these needed grubby research into individual purchasing habits.

Economics has begun to wonder about another nineteenth-century bias it has long retained — the one about 'usefulness'. Making steel is of course 'very useful' — but making music is not. For many years, econo-mists preferred goods and services to meet what they perceive as practical needs. That's why, following the Second World War, economic theory became deeply wedded to the idea that higher con-sumption corresponds to increased well-being — that the higher the 'gross national product' (the measurement economists use of 'all the money' that people in a country spend, one way or another) — the better.

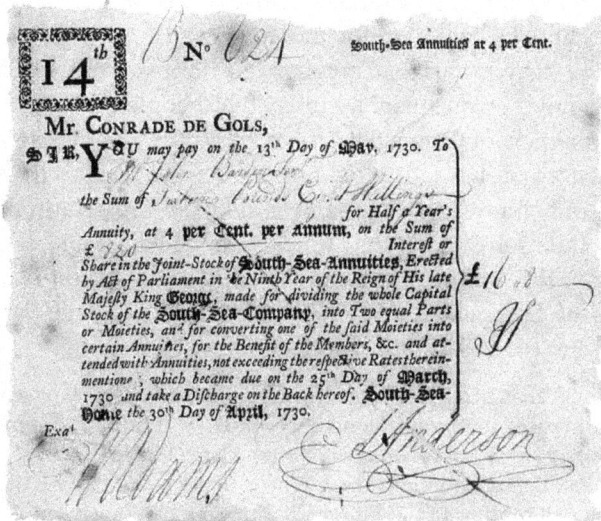

Figure 9. *An excellent investment.* Annuities advance notice promised by the soon to be infamous South Sea Company, 30 April 1730.

Then, in his 1958 book *The Affluent Society* (or, to give its full title, *In Praise of the Consumer Critic: Economics and The Affluent Society Consumption in Mainstream Economics*), J.K. Galbraith says that it is all, in fact, quite the other way around. Galbraith argues that consumers' desires are no longer urgent, or intrinsic, because once a society is affluent, urgent needs have been met, and wants are created by the sales and marketing efforts of corporations. Advertising creates the craving for faster and more powerful cars, bigger and grander houses, more exotic food, more erotic movies — more books even! It is advertising that creates, as he puts it, 'the entire modern range of sensuous, edifying and lethal desires'. And yet, economists, Galbraith says, 'have closed their eyes (and ears) to the most obtrusive of all economic phenomena, namely modern want creation'.

Worse still, he warns, the emphasis on private consumption has crowded out public goods — the public interest. Galbraith is especially prescient on the natural environment, the great loser in conventional economic calculations.

Today standard economics assumes societies are made up of lots of independent individuals — not groups, let alone classes. Yet of course, as customer research specialists know only too well, people fit into very easily identified and surprisingly rigid categories. The man who buys a Ford Escort car reads a certain newspaper, watches certain programmes on a certain kind of TV, and of course votes a certain way. These are statistical trends. Exceptions are anticipated. But economics is about averaging out all the individual choices, and the idea is that patterns ought to be central to the theories.

According to contrarian Canadian economist Tom Green, mainstream economics erroneously assumes that people continuously act in a rational manner, omitting the affects of culture, advertisement, and other influences on human decision making. He accuses conventional economists of being WASP (White, Anglo-Saxon Protestant) males who contradict themselves all the time. (Probably Green is thinking specifically of the author of the most widely read economics textbook, *Foundations of Economic Analysis*, Harvard alumni, Paul Samuelson.) Green parodies the WASP approach, saying:

> '...he admits that utility is a construct that has no basis in psychology; although he uses the terms "consumer" and "individual," his model is built around a fictional character that critics have dubbed Homo economicus. This economic man (yes, he is male) never had a childhood, never has children, has never depended upon a caregiver and does not have anyone he provides care for. He only experiences well-being by consuming. He is rational, selfish, a psychopath... he isn't influenced by hundreds of billions of dollars in advertising or the purchases of his neighbours. If *Homo economicus* buys something, it gives him utility; his consumer sovereignty must be respected.'

The fundamental problem, though, is that economics is about complex relationships with a lot of unpredictable feedback effects built in, of which human psychology is just one example. In fact, predicting economic patterns is more like weather forecasting. That's why, in economics, as with science in general, it is pretty hard to get going without assuming that:

- Simple systems behave in simple ways
- Complex behaviour must have a complex cause
- Different systems follow different rules

Unfortunately, useful or not, and despite being pretty much universally believed, these three traditional Rules of Science are just human inventions. The world is not like that. Indeed, chaos theory reminds people that the truth is very different.

- Simple systems can behave in complex ways
- Complex behaviour can arise in simple systems
- Different systems can follow the same rules

Put another way, there are no rules! Hence the name 'chaos theory'. An orderly system, for example (like the buying and selling of shares on Wall Street), can suddenly and unpredictably become disordered, or change state, without any change in circumstances. It's alarming, and dashed inconvenient too. No wonder not only the experts but the rest of us too try to impose order on the world, to find those elusive eternal and absolute truths Plato promised lived in the World of the Forms. 'It's funny', explained one Quant fund manager (Theodore R. Aronson, whose own funds had just dropped by 40%) to a newspaper, 'but when quants do well, they all call themselves brilliant, but when things don't go well, they whine and call it an "anomalous market".'

The humdrum fact is, in many real-world situations, tiny factors multiply themselves, causing profound changes. That's why chaos theory itself can apply to biology, weather, and even the stock markets. In geological time, the magnetic poles of the Earth itself are known to have 'flipped', changing at a stroke the Earth's climate and weather systems. Yet this massive change can be triggered by no more than a random electrical spark, or, in the world of stocks and shares, by one mistyped sell instruction.

Nicholas Taleb, a financier-turned-academic, and the author of a book on randomness, places the periodic surges and inevitable collapses of the stock markets in this kind of fundamentally unpredictable category. If, every day in the newspapers, movements on the markets are explained 'post-hoc' by experts, that does not mean that their explanations are correct, and (more to the point) that the attached advice and predictions are either.

So why are economic analysts and experts so highly paid and highly revered? For, indeed, this incorrigible faith in patterns powers many high-octane analyst positions in the investment banks and government departments. Technical analysis is as impressive as it is essentially bogus. How many times have you read the explanation for a market fall by someone commenting after the event? It's easy to explain things that way. Predictions, however, are more tricky.

But the experts won't admit that. According to a story told by the analyst Martin Zweig, one of the world's richest investors was once asked at a dinner party by the woman sitting next to him about the 'secret' of his investing success. (At these dinner parties, the investors are indeed WASP men, and the women are just there to ask such questions.) The wise man paused momentarily from his eating and then replied very shortly: 'The secret is, I am rational.'

Taleb, the poacher-turned-gamekeeper financier, sees the motivations of investors rather differently:

'We love the tangible, the confirmation, the palpable, the real, the visible, the concrete, the known, the seen… the salient, the stereotypical, the moving, the theatrical, the romanced, the cosmetic, the official, the scholarly-sounding verbiage… the pompous Gaussian economist, the mathematicized crap, the pomp, the Academie Francaise, Harvard Business School, the Nobel Prize, dark business suits with white shirts and Ferragamo ties, the moving discourse, and the lurid. Most of all we favour the narrated…'

The financial crisis of 2010 cruelly stripped away the grand claims of the economic analysts not only to understanding the innermost workings of the markets, but, which is rather worse, to having a grip on the basics of everyday pricing, which is the essential attribute of being an economist. As Antonio Borges, a former bigwig at Goldman Sachs, now poacher-turned-gamekeeper at the International Monetary Fund, put it: 'At the core of these functions is the ability to find and set the right price, including the extent to which it reflects the risk. This is not really a question of financial sophistication, of complex products or greedy bankers. It is a question of getting the prices wrong.'

Would a truly rational society shut the financial centres all down, before the next disaster? But the 'bottom line' is, Nicholas Taleb explains, that:

'Alas, we are not manufactured, in our current edition of the human race, to understand abstract matters — we need context. Randomness and uncertainty are abstractions. We respect what has happened, ignoring what could have happened.'

Chapter Ten

Climate Science and the Profits of Doom

And ceaseless efforts to predict the future using raw computer power. Illustrating the phenomenon of information cascades

Subject: Climatology

What we're supposed to think:

'You only have to look at Mars, Earth and Venus. Mars has least CO_2 and is coldest and Venus has most and is hottest.'[1]
—Prof Bob Watson (Chief scientific adviser to the UK's Department of Farming), commenting during the *Guardian ClimateGate* 'Debate' held at the RIBA London in 2010

What you're not to say:

'Compared to solar magnetic fields, however, the carbon dioxide production has as much influence on climate as a flea has on the weight of an elephant...'
—Dr Oliver K. Manuel, Professor of Nuclear Chemistry, University of Missouri, Rolla

---

[1]   Actually, the (thin) atmosphere of Mars is 95% CO_2.

Is belief in global warming science another example of the 'Madness of Crowds'? That strange but powerful social phenomenon, first described by Charles Mackay in 1841, turns a widely shared prejudice into an 'authority'. Could it indeed represent the final triumph of irrationality?

Naturally, the debate about anthropogenic (man-made) global warming gets very heated. The theory is that people burning fossil fuels such as coal and oil have changed the amount of carbon dioxide in the atmosphere, preventing the heat the Earth absorbs in the day from escaping at night... John Holdren, the University of California physicist, who said some years ago now that it was possible that carbon dioxide-induced famine could kill as many as a billion people before the year 2020 is an example of the 'Revenge of Gaia', a kind of theological-moral perspective. He said that in the 1980s, when 2020 sounded like a very distant date and certainly quite beyond anyone's ability to check.

So what's the backstory, as journalists used to say before the internet made such research irrelevant? In fact, like many current areas of scientific controversy, when you look into it, it turns out that the debate is a pretty old one. There's nothing new under the too hot Sun! The theory that carbon dioxide would cause the Earth to overheat was first raised in the nineteenth century, but fell out of fashion because it was noticed that any effect it had must be tiny. Out of fashion, but not forgotten. In 1918, a Swedish (Nobel prize-winning, actually) chemist called Svante Arrhenius, having observed through his telescope that Venus was totally obscured by clouds, was confident to write in a book called *The Destinies of the Stars* that a 'very great part of the surface of Venus is no doubt covered with swamps', with humid conditions not unlike the tropical rain forests of the Congo. In a bold sweep of the pen, Venus thus became, for much of the twentieth century, a place for science fiction films and writers to place all manner of unusual life forms, from galactic-dinosaurs to super-intelligent carnivorous plants. Learned comparisons often compared the planet to Earth in the Carboniferous period. But years later, better technology began to reveal a rather less hospitable planet. Observations using spectrometers revealed an atmosphere consisting not of water vapour but almost entirely of carbon dioxide. And the planet was much hotter than previously thought. Hundreds of degrees centigrade hotter. Too hot even for dinosaurs.

Shame! Thus it was that the June 1982 issue of *Popular Science* magazine trailed dire warnings of the effect of having too much CO_2 in the atmosphere, explaining that:

'Venus once had as much water as Earth. It lost the equivalent of Earth's oceans in the process of becoming a runaway greenhouse...'

A block of lead placed inside this greenhouse would turn into a puddle. No wonder that Venus's beautiful seas long ago boiled away... But could a similar thing happen here, the magazine asked? Ah, 'the scenario is complex but seems to fit observations', came the reply from the scientists. One farsighted fellow warned that 'the amount of carbon dioxide we're putting into Earth's atmosphere today is the most dangerous of all human activities...'

The early 1980s was a time when many magazines had only just ceased to predict an icy end to civilization in a ten-thousand-year Ice Age. Almost incredibly, in just twenty or thirty years, a completely opposed view became everyday commonsense with all the 'experts' and certainly all the national governments saying exactly the same thing.

Can We Still Believe What We Think?

It is not just the Vulcans of Start Trek and the philosophers of Oxford who pride themselves on being 'logical'. When it comes to serious matters it seems that everyone likes to think that their beliefs are chosen from among alternatives on the basis of the evidence for them. Yet research by Deanna Kuhn (no relation to the more famous 'Thomas') and others leads fairly inexorably to the conclusion that individuals rarely do hold their beliefs on the basis of evidence but rather arrive at them as a result of social pressures.

Rather ironically, argues Professor Kuhn, as a society we spend much of our time and effort determining what it is that we believe we know, and seem to care little about how it is we come to believe what we do. Kuhn asks: Do people know why they believe what they do, in a way that they can justify to themselves and others? Do they even know what they believe, in the sense of being consciously aware of these beliefs as choices they have made among many different beliefs they might hold? Do they understand what sort of evidence would indicate that a belief should be modified or abandoned?

For Deanna Kuhn, the answer is rather alarming. It seems that many people do not or cannot give adequate evidence for the beliefs they hold. Worse! People are unwilling or unable to consider revising their beliefs when presented with evidence against them. Kuhn holds that reasoned argument requires, at the very least, this ability to distinguish between the theoretical framework and the physical evidence.

At which point, a little scepticism becomes very much in order. Especially considering the policy changes the new theory required. There was to be a shift of world agriculture away from growing food —

to growing biofuels. The European Union passed a law to that effect. There was to be a systematic taxing of heat and light – aiming particularly at people in old houses in colder regions. Meanwhile, nuclear power, for long the bogey man of environmentalists, was reinvented as the planet's saviour.[2] In its Climate Change Act of 2008 the UK government announced a range of subsidies for energy generation that allowed nuclear electricity to be more than twice as expensive than its fossil fuel rivals – and yet still win the best contracts. The strategy was unprecedented – but the costs were left murky and 'in the future'. The government estimated it at a mere £760 or so per household, although you would have had to have been a very optimistic economist to think you really understand the consequences of the new policy for the public finances. The *London Guardian* newspaper, used to warning that the poor are

> '...wilting under the burden of other welfare cuts: changes to local housing allowance, the bedroom tax, the benefit cap and the below-inflation up-rating of benefits and tax credits. A third of households who claimed the old council tax benefit contained children. Many will have no savings as a shock absorber, many others will already have debts'

had no problem with this new charge, representing a sixth of the disposable income of poorer households. All of this was offered as the sane, rational response to the threat identified by climate science. Voices challenging the new orthodoxy were dismissed as irrational and ignorant.

Yet, how rational is it to worry about the effects of climate change on rainforest, and impose targets requiring massive clearance of rainforest in order to grow biofuels? To pass laws banning lightbulbs (maybe urging their replacement by ones filled with poisonous mercury vapour) in order to 'save electricity' – while ploughing public money into schemes to run cars on... electricity? For that matter, how smart is it to pay the Russians once for fossil fuels, and a second time for permission notes (via carbon credits) to burn it? When the Kremlin signed up to the Kyoto treaty it was given an annual emissions limit based on the dirty old Soviet industries – with the result that it has

---

2   As John Ritch, director general of the World Nuclear Association, put it (in June 2007): 'Greenhouse gas emissions, if continued at the present massive scale, will yield consequences that are – quite literally – apocalyptic... If these predictions hold true, the combined effect would be the death of not just millions but of billions of people – and the destruction of much of civilisation on all continents.'

accumulated emissions permits for about 4 billion tonnes of CO_2. Call it a £50 billion political sweetener courtesy of Western consumers.

And how logical is it to suppose that the effects of increased CO_2 in the atmosphere take between 200 and 1000 years to be felt, but that solutions can take effect almost instantaneously?

The Slow but Inexorable Rise of CO_2

Since pre-industrial times, the amount of carbon dioxide in the atmosphere has increased by 40 percent. Half of that increase has come since 1970. April 2014 was the first month that carbon dioxide concentration levels in the atmosphere went above 400ppm.

This is the highest concentration of CO_2 the atmosphere has held in at least 800,000 years and more likely in 20 million years. CO_2 emissions are exceeding levels that would have provided a 50-50 chance of holding the increase in global average temperatures to about 2 degrees Celsius by 2100.

—Scary statistics from the *Christian Science Monitor*.

Whether rational or not, it's certainly not 'science' as we are usually expected to admire it. Instead, climate science in general, and 'global warming' theory in particular, became a compulsory political orthodoxy. A newspaper columnist for *The Boston Globe* announced:

> 'I would like to say we're at a point where global warming is impossible to deny. Global warming deniers are now on a par with Holocaust deniers, although one denies the past and the other denies the present and future.'

Similar statements from people highly placed in government, international organizations, the press, academia, and science made the debate seem closed and the conclusion beyond dispute. Yet the plain fact was that there always was something deeply unscientific about the theory of global warming.

Indisputably, much of what is presented as hard scientific evidence for the theory of man-made global warming has changed and mutated. Sceptics might even insist that calling the scientific account a 'second-rate myth' would be a better term, as the philosopher Paul Feyerabend called science in his 1975 polemic, *Against Method*. Yet a more generous interpretation is that climate science is an extremely complex issue, involving a range of disciplines beyond the ability of any one individual to master, and scientists are only following the longest, if not strictly speaking the best, traditions of their kind by posing hypotheses and attempting to find evidence to support them.

Forget the bearded scientists heroically digging out ice cores in the Antarctic. Climate science is about impressively powerful computers that are fed vast amounts of facts, programmed with a range of rules for interpreting them, churning through incomprehensibly large numbers of calculations before spewing out predictions on the future. The spirit is that of the French philosopher Pierre Laplace, the one who predicted very firmly that one day a great intelligence (a super-super-supercomputer) would be able to calculate the movements of everything in the universe and thus predict the future.

> 'Such an intelligence would be able to embrace in the same formula the movements of greatest bodies of the universe and those of the lightest atom, of it nothing would be uncertain and the future, as the past, would be present to its eyes.'

The Climate Modellers

Climate change theory is built on scientific models of the world, and such a method is certainly splendidly 'detached', yet as even some of the top climate change modellers have remarked: 'Modellers have an inbuilt bias towards forced climate change because the causes and effect are clear.' What the authors of this mean is that a model of climate that produces a 'cause and effect' relationship is preferred to one that produces seemingly random and unpredictable behaviour. (The quotation comes from the paper 'General Circulation Modelling of Holocene Climate Variability', by Gavin Schmidt, Drew Shindell, Ron Miller, Michael Mann and David Rind, published in *Quaternary Science Review* in 2004.)

But reality does not have to fit our rules. A paper by Richard Lindzen and Yong-Sang Choi, called 'On the Determination of Climate Feedbacks from ERBE Data', published in July 2009 in the journal *Geophysical Research Letters*, examined the modellers' case for CO_2-induced global warming. It offered 12 graphs, 11 of them based on the most sophisticated climate models, all but one of which showed that as the temperature of the surface of the seas increases slightly, the amount of heat then trapped in the atmosphere by water vapour increases—a key element in accelerating the 'greenhouse effect'.

Yet there was that odd graph out, the twelfth one. The graph that contradicted all the others was the one based not on a model but on satellite measurements. It showed the Earth's oceans dampening the heating effect.

Certainly, the task of predicting global warming came as manna from heaven for the supercomputer people. Millions upon millions of measurements of air, sea, and ground temperatures, wind speed, tides, dust levels, sunspot patterns, rainfall—et cetera, et cetera—are fed into their expensive machines. Then the computer applies the 'rules', which are grandly declared to be (like Newton's ones) accurate descriptions of how the world works. It all cost fabulous amounts of money. In the UK alone, the government allocated the Hadley Centre for Climate Prediction and Research near Exeter, opened by Mrs Thatcher, a snug £180,293,252 (well over $250 million) for the computer modellers to predict drought in traditionally wet parts of the county — in a hundred years' time. But weather, and especially climate, is not so easily tamed. It is the epitome of a complex system accountable to no one.

History, as ever, gives an insight into the climate change debate. The historian of science, Bernie Lewin, recalls the views of the British scientist who for many years struggled to persuade governments that actually, yes, climate did change. Hubert Lamb, an academic and the founder of the influential Climatic Research Unit (better known by its acronym 'CRU') at the University of East Anglia in the UK, is conventionally credited with putting man-made climate change on the world agenda.

Lamb was described by one of his successors at the CRU, Trevor Davies, probably self-servingly, as 'the greatest climatologist of his time'. Davies credits him with 'convincing the remaining doubters of the reality of climate variation on time-scales of decades and centuries', and an obituary in *Nature* offers his great achievement as overturning the 'old orthodoxy' of climate stability. Other scientific admirers suggest that it was Lamb who first introduced the idea that climatic change has happened, and is still happening, on human timescales. But (as Bernie Lewin notes) Lamb was far from the first to introduce the idea of a constantly changing climate. And Lamb maintained a guarded attitude to the importance of greenhouse warming, Lewin sees nothing odd in the position of a scientist advocating the idea of *natural* climate change being 'guarded' about the evidence of a global *human* influence. Awareness of past variability would rather tend towards scepticism of claims to have identified a single, new and extraordinary cause of climate change.

Hubert Lamb, in fact, was an old-school meteorologist, and there was something of a clash of cultures between people like him, brought up on geology and weather records, and the new kinds of 'climate scientists' only recently tempted by research money into applying their mathematical skills to the natural world. Lamb expressed open scepticism of the theoretical physics used now to predict future climate trends, noting that the models often failed to match reality. He argued for the existence of negative feedbacks dampening warming trends where the climate modellers allowed only positive feedbacks amplifying them; he argued that twentieth-century climate variation was better explained by natural factors (such as solar and volcanic effects), and he wondered how accurate figures relied on for identifying relatively tiny temperature trends were.

Such doubts led him to spend time analysing the political impetus for the new climate science, including vested interests, and the reasons why climate modelling in particular received so much support. Lamb completed his memoirs in 1997, just a few months before Kyoto, the international summit in which man-made climate change was given a pre-eminent role. In these memoirs the man who launched the anthropogenic global warming ship laments:

'It is unfortunate that studies produced nowadays treat these and other matters related to changes of climate as if they are always, and only, attributable to the activities of Man and side-effects on the climate.'

Another Temperature Record Crashes?

In 2015, NASA's Goddard Institute for Space Studies (the main media source for 'facts' on climate change) claimed its analysis of world temperatures showed '2014 was the warmest year on record.' But it soon emerged that its analysis was subject to a margin of error far outweighing the supposed increase. This led the leader of the NASA climate scientists, Gavin Schmidt, to 'precise' their claim as having only 38 per cent certainty. That still sounds grand, and obscures the reality that anything less than 50% certainty means it probably is not the case. Yet the NASA press release failed to mention this qualification, or the fact that the alleged 'record' amounted to an increase over the previous 'warmest ever year', in 2010, of just 0.02°C—or two-hundredths of one degree. The margin of error was approximately 0.1°C—five times as much.

Lamb describes what we might call a small, localized 'paradigm shift' too, when he notes:

'Since my retirement from the directorship of the Climatic Research Unit there have been changes there... My immediate successor, Professor Tom Wigley, was chiefly interested in the prospects of world climate being changed as a result of human activities... After only a few years almost all the work on historical reconstruction of past climate and weather situations, which had first made the Unit well known, was abandoned.'

The paradigm abandoned here is from natural factors producing complex variation, to a new, 'linear' model of greenhouse gases being released into the atmosphere by human activities resulting in a slow but steady increase in global temperatures. Lamb speculated openly about what had caused the science to go astray and what he says provides a cautionary lesson for not merely climate science but for 'big science' in general. A key factor he identified was the distorting influence of *public controversy*.

'Money to fund research may be more or less readily forthcoming according to what the results appear (or are expected) to indicate. This irrelevant influence – to which all countries seem liable in only varying degrees – may be backed by powerful interests and threatens to cloud the possibilities of scientific understanding.'

Additionally, there was the problem of 'fashions in scientific work'. This is, of course, that notion of paradigm shifts, seen in sharp relief.

'...whereby some theory catches on and gains a wide following, and while that situation reigns, most workers aim their efforts to following the logic of the theory and its applications, and tend to be oblivious to things that do not quite fit. The swings of fashion among meteorological and climatic research leaders over the carbon dioxide effect provide an extreme example.'

In his reflections elsewhere on scientific fashion, Lamb was struck by how solar forcing suddenly went out of fashion in the 1930s after bold forecasts based on the sunspot cycle by senior British meteorologists turned out to be wildly wrong. Years later, recalled Lamb, and despite new evidence, for a young scientist to entertain any statement of sun-weather relationships 'was to brand oneself as a crank'. But it was observations of how the fashion for the CO_2 effect came and went as the climate in mid-Northern latitudes warmed and cooled – yet with some years lag – that provided for Lamb the 'extreme example' of scientific fashions. Interest in the relationship waxed mid-century, following early tenwtieth-century warming, only to wane in the 1960s when it seemed obvious that the climate in the Northern hemisphere was getting colder (despite greatly increased carbon dioxide emissions), a cooling that persisted from the late 1950s till about 1974. Finally, after a long run of mild winters in the Northern hemisphere during the 1970s and 1980s, the theory of man-made global warming really took off. Articles on global warming began to dominate the research literature, even if overall temperature averages in some regions, particularly in the Arctic, were still moving downward. Seeing this, with the perspective of someone now in the last years of his life, Lamb complained that:

'...the prospects of global warming are now spoken of on every side and are treated by many, including people whose decisions affect millions, as if the more alarming forecasts were already established as fact.'

The climate scientists who had previously lauded Lamb as a founder and visionary became increasingly bemused and annoyed by his scepticism. Former colleagues at the Climate Research Unit spoke of a generation gap between Lamb and the young researchers now employed. As his successor, Tom Wigley, put it, 'the field moved on, but Hubert did not.'

However, if Lamb was the black sheep in East Anglia, he was not alone among leaders of climatological research of the 1970s, many of whom refused to 'move on'. Robert White, one of the most prominent members of the World Meteorological Organization during the 1970s, and the chair of something called the World Climate Conference in 1979, warned of an 'inverted pyramid' with minimal evidence at the bottom supporting the calls for monumental changes in public policy at the top. William Nierenberg, the director of the Scripps Institution of Oceanography from 1965 to 1986, was another outspoken sceptic. Reid Bryson, the founder of the Center for Climatic Research, CRU's sister organization in the USA, was convinced only of the Earth's tendency towards Ice Ages. Hendrik Tennekes, the director of the Royal Netherlands Meteorological Institute from 1977 to 1990, also spoke out against the new science. And in Australia, as head of CSIRO Division of Atmospheric Physics, Brian Tucker, who had led the research into greenhouse warming from the late 1970s, came out as a sceptic – upon retirement!

Yet the changing of the guard during the 1980s is perhaps best symbolized by Robert Jastrow, the founding director of NASA's Goddard Institute of Space Studies, always sceptical, who handed over to James Hansen in 1981. Hansen was an unabashed alarmist, and that very summer won the first front page headline for global warming in the *New York Times*. It reported how Hansen's research predicted a global warming of 'almost unprecedented magnitude' with potential collapse of the West Antarctic ice sheet, sea level rise, coastal flooding, and widespread disruption of agriculture.

The political uses of the theory of man-made global warming – for new energy taxes, for new carbon exchange markets – revolutionized the status of climate science. With almost bottomless funds available everywhere, is it any wonder that a conceptual gulf between the older generation of independent or retired researchers and the new guard of young, salaried scientists arose?

The gulf is amplified by the preference of the media for televisual new scientists over dreary old academics. The story of how the BBC abandoned its statutory duty to report impartiality is a case in point.

Impartiality, Experts, and the BBC

All BBC staff carry an identity card which proclaims the first BBC's mission to be independent, impartial, and honest yet, as a recent (2007) BBC report, gnomically entitled 'From Seesaw to Wagon Wheel' (subtitled a little more helpfully: 'Safeguarding Impartiality in the 21st Century') acknowledges, impartiality is easier to talk about than actually achieve.

As the BBC report says, 'Recent history is littered with examples of where the mainstream has moved away from the prevailing consensus.' Take that big issue of reporting macroeconomics and government policy. In the 1970s, monetarism was 'an eccentric, impractical enthusiasm of right-wing economists', but a generation later it was a central feature of every European government's economic policy.

On the other hand, over the same period, attitudes in much of Europe towards the European Union had turned Euro-scepticism from what 'was once belittled as a small-minded, blinkered view of extremists on both left and right' into 'a powerful and influential force'.

The report noted that climate change is another subject where dissenters can be unpopular, adding:

'There may be now a broad scientific consensus that climate change is definitely happening, and that it is at least predominantly man-made. But the second part of that consensus still has some intelligent and articulate opponents, even if a small minority.'

Actually, the BBC's report says that the Corporation's policy on the huge issue of climate change should be that dissenters' voices should still be heard, because, one, it is not the BBC's role to close down this debate, and secondly because impartiality 'always requires a breadth of view' and that 'bias by elimination is as dangerous today as it ever was'. It even adds that the BBC has many public purposes of both ambition and merit—'but joining campaigns to save the planet is not one of them.' Instead, programme-makers should reflect the full range of debate that such topics offer, scientifically, politically, and ethically.

However, the alternative to reflecting a range of views is to try to find the 'consensus' view and that is the direction the organization went in. Alas, finding the consensus view is by no means as easy as many people imagine.

But how do you obtain a 'balanced view' on complicated, not to say highly political, matters? The BBC started off by holding, they said, a high-level seminar in Exeter, in South-West England, 'with some of the best scientific experts'. After this, it came 'to the view that the weight of evidence no longer justifies equal space being given to the opponents of the consensus'. Where once news reports and documentaries might have said that many scientists believe that burning coal and oil is not only causing the ice caps to disappear (cue pictures of drowning polar bears in the *Daily Mail*) and even the towering Himalayas to melt, now the BBC simply stated such things as fact.

Researching the experts' views sounds a good way to approach the matter. But you don't have to be a rocket scientist to know that different experts have different views—and some experts are not quite as neutral as they claim to be. Exeter, it is not irrelevant to mention, is a long way from BBC HQ—but right at the heart of the UK government's Met Office's research effort into demonstrating man's influence on the climate.

> So how were the BBC experts selected? People asked this question and the BBC went to court to prevent the information ever becoming public, to keep, as they put it, their 'sources' secret. The UK courts agreed.
>
> Just unfortunate then for the BBC, that an Italian climate sceptic found all the names of the people at the expert conference still there on a long-forgotten web-page. And the list showed that far from a representative sample of scientific opinion, the meeting consisted of special interest groups with financial interests, such as British Petroleum (oil companies, despite what you may have read many times, are one of the big winners of the 'coal is bad for you' policy as (for example) they own most of the world's gas reserves), plus scientists whose jobs revolved around proving the theory of man-made climate change and, of course, plenty of green campaigners whose commitment to the cause of fighting global warming was in inverse proportion to their expert knowledge.
>
> These experts' evidence, however well intentioned, was all bound to come to one conclusion. The desire to keep the participants' names secret revealed that the BBC knew it was on very weak ground. But that's how, despite the famous card in every staff members' pockets proclaiming and promising impartiality above all, the BBC adopted a policy that turned reporting about climate issues into propaganda for a particular cause.

How this situation came about says much about how science is co-opted to sway public opinion. The case is built, deliberately or not, on misleading images and interpretations that have been perpetuated by parties with a vested interest. It morphs into a tool for governments to intimidate their populations into passive acceptance of very real changes: from the tiny, such as accepting miserable fluorescent light instead of the incandescent light we've been used to; to the major, like welcoming nuclear power plants and obliging rainforest tribes to make way for biofuel plantations. Indeed, much of what is presented as hard scientific evidence for the theory of global warming is false. Recall that accusation of science as 'Second-rate myth' made by Paul Feyerabend in his 1975 polemic, *Against Method*. 'This myth is a complex explanatory system that contains numerous auxiliary hypotheses designed to cover special cases, as it easily achieves a high degree of confirmation on the basis of observation', Feyerabend wrote.

> 'It has been taught for a long time; its content is enforced by fear, prejudice and ignorance, as well as by a jealous and cruel priesthood. Its ideas penetrate the most common idiom, infect all modes of thinking and many decisions which mean a great deal in human life...'

The important thing, myth or science, call it what you will, is that you don't think that by calling it 'science' it becomes irrefutable. Because that it ain't. Even in the flexible world of climate science, there's no such thing as an 'average global temperature', nor even an average global CO_2 level. The figures you arrive at depend on what you choose to measure, where you choose to measure it, and at what time. The selection of data is all.

Nor are there any experts in it — only various kinds of amateur. This is because the Earth's climate is an enormously complex subject, spanning not only the 'pure' sciences like physics and chemistry, but many of the 'natural sciences', such as oceanography, meteorology, volcanology, palaeontology, archaeology, solar science, and many others.

All this makes becoming a highly skilled 'Climate Scientist' very challenging. So step forward mathematician Michael Mann with his hockey-stick chart overturning long established notions of global climate trends, accepted virtually overnight without close examination. Shortly followed by Al Gore, who despite near-failing grades in science and math in college has a long-standing interest in 'the environment', and who decides to make a movie out of it.

Gore's presentation soon became one of the most popular works arguing the case for global warming and the need for action. In *An Inconvenient Truth*, the film about Al Gore's campaign to educate people about global warming, the scientists are reduced to a walk-on part: they are in essence an audience invited to applaud the decisions of politicians. The former Vice-President unveils as the 'scientific' highlight of his presentation a graph offering a clear correlation between CO_2 and temperature, as discovered in core samples of polar ice at one particular location. He goes on to state that as levels of carbon dioxide rise, the Earth's temperature increases because the atmosphere traps more heat from the Sun. Driving his point home, Gore extends the lines on the graph to terrifying, if distorted, levels.

There are many ways to fool people, and linking images with complex theories is a good one. Another good way, of course, is simply to add misleading pictures. And indeed, as part of its effort to show that the effects of carbon dioxide are already being felt, the film presents striking images of 'global warming', from forlorn boats in dried-up seas to that haunting image at the end of the film of polar bears clinging desperately to a shrinking block of ice.

The film — like the theory it was advancing — is not defensible either in terms of factual accuracy or of argumentative logic. That sobering image of fishing boats stranded in a desert that was once the world's largest freshwater sea (which features not only in Al Gore's *An Inconvenient Truth* but in his 1992 book, *Earth in the Balance: Ecology and the Human Spirit*) is misleading because the Aral Sea is actually not a sea. It is rather a huge lake dependent on constant inflows from rivers, which have been gradually choked off since the end of the Second

World War by Soviet-era irrigation schemes. Its plight has nothing to do with global warming.

All right, but what about the polar bears then? I can't help but feel sorry for them. So cuddly! But it turns out that when Gore used a picture of two polar bears supposedly stranded on a melting iceberg to support his claims about global warming, he chose a photo that maybe he shouldn't have done. It was one that had been taken by Amanda Byrd, a marine biology student, on a research cruise in August 2004, a time of year when the fringe of the Arctic ice cap normally melts. The image was later distributed by Environment Canada, a Canadian government department, to media agencies. With that polar bear picture on the screen behind him, Gore says, 'Their habitat is melting... beautiful animals, literally being forced off the planet. They're in trouble, got nowhere else to go.'

Perhaps Al Gore didn't know that polar bears can swim. Perhaps he didn't know that Amanda Byrd stated that, when she took the picture, the bears didn't appear to be in any danger. But he should have done. As an Environment Canada spokesman, Denis Simard, told *The National Post*, a Canadian newspaper, you 'have to keep in mind that the bears aren't in danger at all. It was, if you will, their playground for 15 minutes... they were not that far from the coast, and it was possible for them to swim.' The polar bear is still the symbol of the effects of global warming—but it is a cleverly designed marketing symbol, and not a rational, scientific marker.

But back to the 'scientific' highlight of the film, offered as evidence of a clear correlation between CO_2 and temperature, as is discovered in those ice cores. About 40 minutes into the film, Al unveils that graph correlating temperatures and CO_2 emissions with the words:

> 'The relationship is actually very complicated but there is one relation-ship that is far more powerful than all the others and it is this. When there is more carbon dioxide, the temperature gets warmer, because it traps more heat from the Sun inside.'

Then, like a rather leaden showman, he asks 'Do they go together?' while extending the lines on the graph to terrifying levels. Unfortu-nately you can't do that in this case, any more than you can extrapolate the height a tree will grow to by measuring its growth in the last thirty years. It's just not science. Or even maths. In fact, its not even very good theatre.

Still not sure whether or not to throw out your low energy light bulbs? But let's look at some of the figures for the day's most evil pollutant—carbon dioxide. Now the atmosphere is known to contain about 780 gigatonnes of carbon (whatever a gigatonne is, but it does not matter, it is proportions that we are interested in here) and it is thought that about 90 Gt is exchanged each year with the oceans along with another 120 Gt with plants. It's a cycle, you see, a dynamic system — the carbon cycle that life on the planet depends on. A side effect of nuclear bomb tests has shown that the half-life of carbon dioxide in the atmosphere is less than ten years. But that's not all, the oceans today have about 40,000 Gt of carbon dissolved in them, and the Earth itself contains an estimated 70,000,000 Gt in the carbonate rocks (like chalk, for example) laid down over the aeons. 'Carbon is constantly exchanged between these and other reservoirs, among which the small modern atmospheric reservoir is carbon starved compared with earlier geological history', says one geologist, Robert Carter.

Mankind's contribution thus remains disappearingly small compared to the vast mechanisms of the planet. One estimate, accepted by the IPCC themselves, is that the human production of carbon dioxide is about 7.2 Gt a year. This amount is so modest that it is lost in the natural cycle—which recall is about 200 Gt a year. That of course doesn't stop journalists from pushing claims, for example, that human CO_2 has changed the acidity of the Earth's oceans, as put forward in many colourful magazine articles about the plight of the disappearing coral reefs and so on.

Given this context, how can CO_2 be imagined to be a threat to the planet? The explanation is that the IPCC predicts a tiny change in CO_2 will have an effect multiplied many times over, via changes in such things as water vapour and cloud formation. It is magnifying factors like this that the computer programmers, who are ultimately the 'authorities' behind its dire warnings, have written into their programs.

If CO_2 levels increase slightly, the programs say, clouds will form at certain specific heights and times to trap more of the solar heat. It's never done it in the past, but it will this time. This in turn will cause more clouds, melt the ice caps, kill the forests, and 'hey presto' — runaway greenhouse warming. Yet these are not facts but human guesses, albeit made more impressive by being conveyed via large computers. It is all a fiction, reminiscent of what Joseph Weizenbaum warned us would happen if society allowed itself to fall for the dangerous allure of computer technology.

The sceptical geologist Robert Carter also reminds us of this sinister statement from sociologists at the Institute for Public Policy Research in London, amongst others:

'The task of climate change agencies is not to persuade by rational argument—Instead, we need to work in a more shrewd and contemporary way, using subtle techniques of engagement—The "facts" need to be treated as being so taken-for-granted that they need not be spoken.—

> Ultimately, positive climate behaviours need to be approached in the same way as marketeers approach acts of buying and consuming – it amounts to treating climate-friendly activity as a brand that can be sold.'

As Carter puts it in a book called *Climate: The Counter Consensus* (2010), climate change has the hallmarks of a conspiracy, made worse by the fact that many of the people taken in by it had only the best of intentions, but lack 'the science education to see through the scam'. Carter concludes by pointing at the true costs of the policy – to the 1,500,000,000 underprivileged people of the world, without clean water, adequate sanitation, basic education, and basic health. 'For lifting the poor out of their poverty, and helping them to generate wealth for themselves, is the only sure way to protect Earth's future environment.'

But it's certainly not just Al who has got it wrong. Look at this glowing endorsement of the film by Eric Steig, an 'isotope geochemist', for *RealClimate.org*, a website run 'in their spare time' by employees of the Goddard Institute for Space Studies, part of NASA.

> 'How well does the film handle the science? Admirably, I thought. It is remarkably up to date, with reference to some of the very latest research.'

It is rocket scientists like these who first noted the lessons of Venus for the Earth, of course. But just suppose, unlike the experts of the Nobel Prize Committee who awarded Al Gore their special prize (along with the United Nations Panel tasked with describing the effects of 'greenhouse gas emissions' on the planet), let alone all the Heads of State who signed up to the public declaration that the 'facts are settled'... just suppose you have a spare one minute to consider rival views to those in the film – what then? Even a minute provides a few pause-worthy facts.

Historically, CO_2 levels and temperatures have *not* marched in 'lock step'. Over geological times, the only thing the two variables share is a 'random walk'. What geologists call the Late Ordovician Period saw CO_2 concentrations nearly 12 times higher than today – 4400 ppm – and was also an Ice Age. In fact, you have to search the charts for the last 600 million years to find even two occasions when CO_2 levels have been this low – at below 400 ppm. One is the time of the dinosaurs when the fossil fuel deposits were laid down – the Carboniferous period – and the other period – going back 65 million years! – is our present one.

Even restricting the scale of our survey to the last 100,000 years (up to the last Ice Age, as recorded in those ice cores so impressively dug up by beardy types), the relationship is not as Al and others thought.

Far from increases in CO_2 leading to higher temperatures, the ice core record shows rises in temperatures *preceding* (by between 200 and 1,000 years) increases in CO_2. Which is what you would expect. Almost all of the CO_2 in the atmosphere comes from the seas. When the sea is warm —like a fizzy drink going flat in the Sun—the carbonized water forms atmospheric CO_2. Higher temperatures mean more CO_2—naturally.

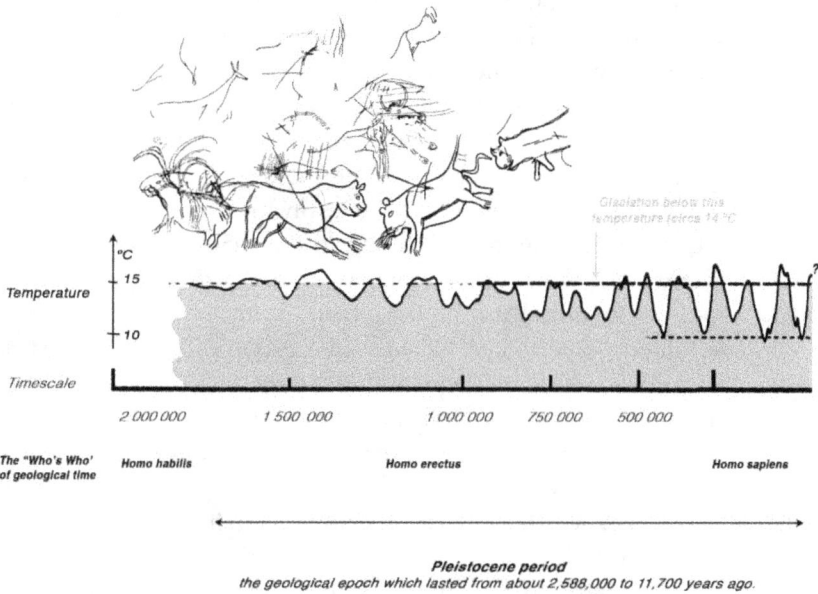

Figure 10. Schematic showing the historical range in temperature on Earth. Top left, the Lascaux cave paintings. The data is drawn largely from geological examination of rocks and sediments for indicators of past condition such as oxygen isotopes or even pollen grains. Such research convincingly demonstrates that the difference between the long-term mean global air temperature and a glacial period is as little as 2°C, and that the difference from the long-term mean to today's warm period is only 2–3°C warmer. In other words, it takes only small changes in temperature to switch global climate into 'ice house' mode. The most recent Ice Age to occur in the Quaternary period (which began 2 million years ago) started around 30,000 years ago, peaked 12,000 years later, and ended between 12,000 and 10,000 years ago. The impact of this Ice Age, which in Britain saw glaciers reach as far south as Bristol and London, is all around us in the land-forms. Geographers know that features such as hanging valleys and U-shaped troughs in the Lake District, erratics and Crag and Tails in Scotland all indicate Britain's recent and much colder glaciated past, as well as evidence of natural climate change. Since then, and starting long before humans influenced climate, the Earth has entered into one of the warmest periods in its history.

What, though, about all the recent *evidence* of global temperatures soaring? Oddly enough, claims of record high temperatures have nothing to do with the theory of anthropogenic global warming. This is because the theory concerns the amount of heat radiated back into space at night – not how much heat is 'getting in' during the day. The theory is interested only in the supposed warming of the surface of the world's seas at night time under a protective cocoon of atmospheric CO_2 – the greenhouse. And is that happening? Er… no. Quite the reverse, or so satellite measurements seem to show.

All this is pretty complicated and no wonder journalists can't be bothered to report any of it. Anyway, that minute must be long ago up. But let's at least agree on the polar bears. They are not in any danger from melting ice caps. Only hunters and oil prospectors spurred on by the higher prices for oil that climate science has also created. Now all right, you may say, but if global warming really boils down to a few media tricks, how come everyone believes in it? Yet, as the American social psychologist Solomon Asch found in the 1950s, people are quite prepared to change their minds on even quite straightforward factual matters in order to 'go along with the crowd'.

In one famous experiment, he showed a group of volunteers cards with various lines drawn on them, and asked them to determine which of the lines were the longest. Unbeknown to one of the group, all the others were not, in fact, volunteers but stooges, previously instructed to assert things that really were obviously not the case, such as that a line that was obviously shorter than another was actually a bit longer… In fact, it turned out that when enough of their companions seemed sure about something, around one third of people were all too prepared to 'change their minds', and (disregarding all the evidence) bend pliantly to peer pressure.

You can't blame folk for doing that. Especially when to do otherwise would mean taking a look at the scientific issues in 'climate change' theory. But actually, it's not as complicated as its proponents, who are essentially mathematicians, like to make out. Like Al's film, we can start by looking at the recent history of the Earth's climate and the most famous image of them all, the so-called 'hockey-stick' curve of the Earth's climate mentioned previously. The chart, correlating temperatures and carbon dioxide levels, appears not once but five times in the United Nation's (IPCC's) landmark 2001 report on global warming, the report which paved the way to ratification of the Kyoto Protocol. It's called the hockey stick as it depicts the Northern hemisphere's temperatures over the last 1,000 years as a fairly regular flat line right up until the late twentieth century when, suddenly, it curves sharply upwards as runaway warming appears.

The graph contradicted all the previous research into climate which had thought there was a warm period around a thousand years ago (during which grapes were recorded being cultivated in Scotland and Nebraska was covered in sand dunes), followed by the 'Little Ice Age' in the fourteenth century, immortalized in oil paintings of the period and contemporary accounts of the great rivers of Europe, such as the Thames, the Seine, and the Rhine, freezing up entirely, with ice parties being held on them.

Jolly Hockey Sticks?

Modern global warming science starts with a remarkable discovery. In this case the finding by Michael Mann, a palaeoclimatologist (one who attempts to interpret the past climate through certain palaeolithic records, such as ice core samples, sea bed sediments, coral heads, and tree ring growth), who submitted a paper to *Nature* journal in 1998. This groundbreaking paper was not even subjected to peer review before publication.

In it, Mann offered a graph showing a dramatic and indeed alarming increase in recent years in temperatures, neatly paired with records of global carbon dioxide levels since around the tenth century. His graph soon became famous as the 'hockey-stick' chart.

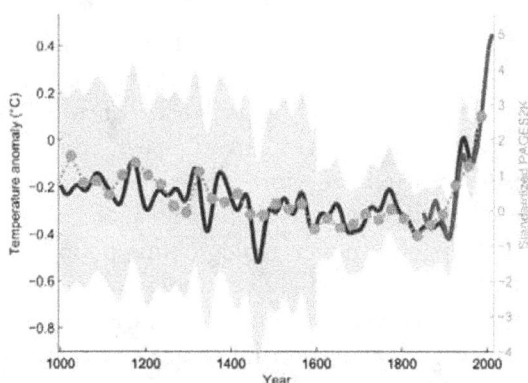

Mann's graph showing the Earth's temperature as relatively constant for the past thousand years before suddenly shooting upward at the dawn of the twentieth century (from: http://en.wikipedia.org/wiki/File:T_comp_61-90.pdf).

The 'interpretation' (rather the message) of the paper was that humans' production of CO_2 was responsible for the sudden increase. There is, however, something odd about the chart.

1. Although it purports to represent 'global' temperatures it in fact only covers the Northern hemisphere.

2. And as everyone—except the editors of *Nature* apparently—knows, during the last thousand years there have been two striking events. The first is the 'Medieval Warm Period' when temperatures, at least in Europe, were significantly higher. The second was the 'Little Ice Age', a period in which the temperatures dropped so low that the Thames in London froze over.

Hardly anyone disputes these climatic events. But the graph does not show them. Why? It turns out that Dr Mann used a statistical method which removed (ironed out) these 'large' temperature fluctuations, leaving the last thousand years as having a completely regular climate, with temperatures ranging only between plus or minus half a degree. But this statistical trick belies reality. In fact, the temperatures ranged much wider. It is a gross example of how statistical manipulation distorts the data. But wait, the chart shows temperatures shooting up at the end—in the last few decades? Ah yes, but these figures have not been 'ironed out'—these are just recent actual measurements which Dr Mann felt needed no 'ironing out'.

This chart became the foundation for the first report of the United Nations International Panel on Climate Change (IPCC), which in turn provided the summary information and recommendations to the world's governments. The lead author of the relevant chapter was very impressed with it. Who was that then? A colleague perhaps? A friend? No, nothing like that. It was… one Michael Mann.

By the time climatologists had pointed out that the graph had somehow missed these well-known (historical) warm periods and Ice Ages alike, and by the time mathematicians had pointed out it had exaggerated the heating at the 'curved' end by bolting unrounded, actual recent measurements onto the much longer series of rounded, estimated figures, the powerful image had already been immortalized in ten thousand news reports and *that* film.

But no matter if, unfortunately, there is no such correlation between CO_2 levels and temperature over geologic time periods. Because more stimulating than that investigation anyway is another puzzle altogether, which is: just what was Al Gore, a US politician and former Vice-President, doing making a documentary about the relationship of CO_2 levels to historical temperatures? This is a much more interesting question and indeed the answer is interesting too. It involves media and marketing experts, people like a shadowy outfit in New York called David Fenton Communications, who exist to promote certain 'clients' by manipulating public awareness of certain issues. Remarkably, Fenton manages to not only be shadowy, but to be the 'world's largest "public interest" PR firm'—not that it likes that term. And 'Communications Director and Environmental Advisor' to Al Gore at the time was one Amy 'Kalee' Kreider, who could always be spotted standing alongside Al whenever he testified on environmental matters to Senate Committees, and whose scientific credentials (Bachelor of Arts, Rollins College, Orlando) seem less relevant than the fact that she was a staffer at Fenton Communications when Al Gore was there getting advice. Other than that, she worked at Ozone Action, where she distributed 'ozone-safe strawberries' to the national press corps.

And so, of course, 'David Fenton Communications' was there at the Kyoto summit in 1997 working with 'green NGOs and leaders' including Al Gore and the United Nations Intergovernmental Panel on Climate Change (IPCC). It was there advising on how to 'mainstream the climate threat' and to 'harness the public "tipping point"' on the issue and inspire action—as its website today boasts. And indeed the public have been well and truly tipped.

The IPCC reports, which should be useful to governments as the basis for their policy discussions, began to say exciting things like 'numerous long-term changes in climate have been observed [including] changes in Arctic temperatures and ice, widespread changes in precipitation amounts, ocean salinity, wind patterns and aspects of extreme weather including droughts, heavy precipitation, heat waves and the intensity of tropical cyclones.'

They even began to warn (amongst other horrifying things) that:

* in just 30 years (that would be by about 2035), all the glaciers in the Himalayas would have melted, leaving one billion people without fresh water;
* over half of the Netherlands would be submerged under the sea;
* up to 40% of the Amazonian basin would die as the rainforest 'could react drastically to even slight variation in precipitation'; and that...
* food production in sub-Saharan Africa would be devastated and millions of people would starve to death.

And although all this was dressed up as science, not a single 'real' specialist could be found to step up to the podium to defend this sort of language, although plenty of helpful green groups and journalists would. On the other hand, and historically significant, scientific bodies did want to sign up. The president of the oldest state-sponsored empirical science academy, Britain's Royal Society, launched it into the new millennium with a declaration for the cause, thus violating an ancient policy 'never to give their opinion as a body'. Similarly, in 2007, the American Physics Society backed up the IPCC's declarations by proclaiming global warming a truth 'incontrovertible'. Since every physicist knows that there is nothing in empirical science beyond challenge, such declarations are not about science but only about dogma. So how did they get into a 'scientific' report—and why?

The explanations are more prosaic and personal than grand and conspiratorial. The claim about the Himalayas was made by a 'glaciologist' employed by the Head of the IPCC (the body in charge of collecting up knowledge on climate science) in his own private research institute. When the figure was challenged, the two men speedily retracted, speaking of having been misquoted by a popular climbing magazine. Yet they used exactly the same claim to obtain a large research grant from the European Union. The claim about the Netherlands is simply based on how much land there is at present below the local sea level. It's highly speculative, and worse, the actual figure for the land below sea level in the Netherlands was 26%. (Indeed, notwithstanding the relentless rise of global CO_2, at the time of writing, it still is.) The claim about the parched rainforest was extracted from a report *on the effects of illegal logging*, as recycled by a green group, the World Wide Fund for Nature. It had nothing to say about the sensitivity of rainforests to future changes in regional rainfall levels.

Yet, given the stakes of saving the entire world, for many green organizations, newspapers, and even governments, such scare stories jolt distracted or myopic publics from their complacency. And increasingly they enjoy professional help in this task, consulting Public Relations firms and polling experts. That world beating 'public interest' PR firm, Fenton Communications, in particular specializes in scary stories of invisible substances causing horrible effects, cloaked in pseudoscientific language. Amongst their campaigns have been the scary effects of an invisible chemical on apples (called Alar) which 'scientists say' may be causing illnesses and killing people.

> 'The most potent cancer-causing agent in our food supply is a substance sprayed on apples to keep them on the trees longer and make them look better. That's the conclusion of a number of scientific experts. And who is most at risk? Children, who may someday develop cancer.'

So reported CBS's flagship current affairs programme *60 Minutes* in 1989 in an exclusive negotiated with David Fenton. Only later did it emerge that the 'scientists' were two campaign activists and that their findings had never made it into any peer-reviewed journal. And of course that they weren't true.

Similarly, the fear that an invisible chemical in milk (the Bovine growth hormone given to cows) that 'scientists believed' might be causing illnesses and killing people, once spread around like so much healthy farmyard manure by a group called Environmental Media Services (who received almost a million dollars in 1999 from a foundation created by Fenton, and themselves set up that website for climate science quoted earlier endorsing Al's film), benefited another Fenton client, Ben and Jerry's Ice Cream. Wholesome, friendly Ben and Jerry's who had made one of their brand's selling points the fact that it was made with hormone-free milk. The public's response to the Alar reports was predictable. 'Panic set in almost overnight. Parents poured apple juice down the drain, and many consumers stopped buying apples and apple products', writes William Kucewicz in a booklet called *The Great Apple Scare: Alar 20 Years Later* (2009).

Not to forget the scary effects of invisible chemicals on plastics (called plasticizers, especially phthalates) which 'scientists say' may be causing illnesses and killing people, especially little children. And now not only green groups but national governments too want help explaining that for decades an invisible gas, CO_2, has been being belched into the atmosphere by filthy industries and millions of cars, and it is already causing worldwide havoc with nature! What a message!

But the big campaign for Fenton was the climate change one. And public interest or no, amongst its clients are all the most respected names in the environmental business from Greenpeace and Friends of the Earth to the UN Environment Program... the list goes on and on! All green groups paying a PR firm to 'sell their message'. It's a big business this, and not surprising, then, that they want to have professional help running the campaign. Even if the results are a little, shall we say, scary. As Bonner Cohen (no relation) wrote in a book on Fenton Communications entitled *The Fear Profiteers*:

> 'The scares just don't ever stop. But they all have one thing in common — a lack of evidence and abundance of deceit. The claims involved in the scares have all been refuted in public. By the time the scares have been debunked, however, the campaigns have taken such a strong hold that the truth usually is irrelevant.'

There has been quite a fashion in the Western world recently (especially amongst scientists and philosophers) for books decrying 'irrationalism', and sneering at the 'sheep-like' followers of religion. Alternative health, astrology, but above all 'climate deniers' have all now firmly been told they must fall into line with 'the science'. But, as is particularly clear with the claims for 'man-made global warming', if this is science, it is also propaganda. And we are not being asked to be more rational but to suspend our own judgment completely. Perhaps that, not 'runaway climate change', is the most dangerous threat to the world today.

Chapter Eleven

The Risk Factor

*featuring reassuring voices on nuclear physics.
Illustrating the political uses of apparently
objective, indeed statistical, arguments*

Subject: Toxicology

What we're supposed to think:

'Ocean currents will disperse radiation particles and so it will be very
diluted by the time it gets consumed by fish and seaweed.'
—Hidehiko Nishiyama, of the Nuclear and Industrial Safety Agency in
Japan

Add to which:

'It's not possible to achieve a climate solution based on existing tech-
nology without a significant reliance on nuclear power.'
—Jason Grumet, President of the Bipartisan Policy Center in
Washington

What you're not to say:

'In the 30 hospitals of our region we find that up to 30% of people who
were in highly radiated areas have physical disorders, including heart
and blood diseases, cancers and respiratory diseases. Nearly one in
three of all the newborn babies have deformities, mostly internal.'
—Alexander Vewremchuk, of the Special Hospital for the Radiological
Protection of the Population in Vilne, Ukraine

If there is one thing most of us want to be able to rely on experts for, it is advice about avoiding dangerous hazards. Is it dangerous to use mobile phones in bed? Will running marathons improve or damage your health? And can you eat free-range eggs or drink unpasteurized milk without getting dangerous bacteria in your stomach? More sophisticated questions arise over whether such and such a food additive is safe, or whether such and such a level of pesticides in drinking water is harmless... or whether that neigbourhood nuclear reactor is quietly putting out toxic by-products.

Such questions are the bread and butter of expert groups whose job it is to produce at least vaguely scientific responses not only for public authorities and regulators but for the concerned readers of Sunday newspapers too. All such questions relate to a particularly modern phenomenon doubtless linked to the decline of popular religion, which is a concern about risk. And as far as health issues go, risk is considered something to be avoided 'at all costs'.

And so, in many areas, governments have moved rapidly to reassure the public. Generally, it seems, the 'modern' state, via its appointed representatives and anointed experts, adopts a very strict line. Where once people could judge risk for themselves, now they are both forbidden to do so, and often unable to do so, as the decisions are taken on their behalf.

The risk of cancer from secondary smoke, in bars or even in the street, has been banished by specific laws banning smoking. The risk of intoxication by over-consumption of alcohol has been reduced by outlawing 'drinking and driving'. Likewise many other risks while driving, such as using a mobile phone, not wearing seat-belts, not carrying a yellow reflective jacket, have been considered sufficiently serious as to require legislation. If in much of the 'developing South' you can often see young teenagers driving around on mopeds with their five-year-old brother or sister perched behind them, in the 'rich North', children the same age as the young bikers must be locked securely into wrap-around padded cocoons.

In the developed world, the risk of side effects from everyday appliances, or from the things we eat, or the drugs we take, has spawned a vast industry of testing (and its dark flip side, vivisection). But conversely, certain things that ordinary people think might be dangerous, the government experts insist are safe. High pesticide levels in water, traces of heavy metals in food, plasticizers everywhere, even invisible microwaves and powerful electromagnetic fields – all have been considered suspect at some time or another, yet all have been forcefully defended by the 'authorities'.

Just as the public remained content to be told that climate science was made by 'climate scientists', even as in reality it was crafted line by line by PR professionals, subcontracted by political animals from Al Gore to the United Nations' IPCC committee, so too the public remains blissfully unaware that often official government pronouncements on environmental health have been shaped by marketing experts acting for special interests.

Take Regester Larkin in the UK, for example. Regester Larkin is a PR company, co-founded by Mike Regester and Judy Larkin. It specializes in 'risk management' for large corporations and government departments alike. Clients and 'partners' include drugs behemoths like AstraZeneca, GlaxoSmithKline, energy giants like Shell, and tobacco companies like the Philip Morris Management Corporation. Companies like this are highly political. No wonder, then, that when the new supposedly 'pro-science' Labour government arrived in Britain in the 1990s, Regester Larkin was immediately put to work as part of New Labour's contract with industry.

Regester Larkin were from the beginning deeply involved in New Labour's contract with industry. One of their first jobs was to work with the Department of Trade and Industry on the perception of the nuclear industry, which was felt to have attracted an unfairly low public status over the years. Of course, no one employs marketing professionals to investigate actual dangers or risks. They are simply employed to reassure people about them.

Judy Larkin is also on the advisory board of another Anglo-American risk management PR company called ECHO, which provides such services to governments and multinational businesses. But PR companies lack a certain credibility. Much better then that Ms Larkin is also on the advisory board for Kings College's Centre for Risk Management. The Centre opened in January 2002. A decade later it had nine academic and research staff, and eleven research students. A sceptic might ask: 'Why does a University Hospital research department need one of the big Anglo-American public relations figures on its advisory board?' For critics like Martin Walker, the answer is simple. If a centre's real business is playing down risk, not researching it, it's ideal that their most noted player would be a PR, crisis management guru.

Clearly what industry wants is not objective scientific research, but a constant stream of disclaiming information that can be channelled out of impressive sounding things like 'Kings through the Science Media Centre' to journalists and politicians.

Indeed the projects at Kings College London have nothing to do with epidemiology or the real measurement of physical illness. The starting point is how people 'perceive' the effect upon themselves of environmental hazards, be it radiation from mobile phones or from nearby nuclear reactors, and the relationship of this to their perception of risk. For many years, to work there under Professor Simon Wessely and a team of clinical psychological researchers involved trying to prove that people who think they have been damaged by environmental factors were really suffering from 'false illness beliefs'. A British government Quango called the Radiation, Risk and Society Advisory Group, chaired by one Professor Sir Kenneth Calman, the former Chief Medical Officer, was an enthusiastic supporter. The Radiation, Risk and Society Advisory Group's work was essentially to help the National Radiological Protection Board in its task of reassuring the public about nuclear matters. As Martin Walker, author and indefatigable campaigner for alternative medical approaches, explains in *'The Lobby, Brave New World of Zero Risk'*, an online pamphlet:

> 'The object of all research is to convince the public that they are involved in acceptable levels of risk. What centres of this kind are measuring is what industry can get away with. And then in a secondary sense how industry can combat bad stories of environmental ill health, from other scientists.'

Indeed, in Britain at least, there has been a significant shift in the public perception of the dangers. Sprouting like so many mushrooms in the British press in the wake of Fukushima, authorities like Tony Brenton, sometime British ambassador to Moscow, wrote that 'nuclear remains one of our safest and cleanest sources of electricity.' The assessment of the *Times* newspaper's science editor was that 'nuclear power is considerably safer than some of the alternatives' and that the important thing was for governments to upgrade to new nuclear reactors.

Or recall Green Guru and Respected Scientist, James Lovelock's words again on the lessons he drew from Chernobyl. They're not entirely what you might expect from a self-proclaimed environmentalist.

> 'Opposition to nuclear energy is based on irrational fear fed by Hollywood-style fiction, the Green lobbies and the media. These fears are unjustified, and nuclear energy from its start in 1952 has proved to be the safest of all energy sources. We must stop fretting over the minute statistical risks of cancer from chemicals or radiation. Nearly one third of us will die of cancer anyway, mainly because we breathe air laden with that all pervasive carcinogen, oxygen. If we fail to concentrate our minds on the real danger, which is global warming, we may die even sooner, as did more than 20,000 unfortunates from overheating in Europe last summer.' ('Nuclear Power is the Only Green Solution', for the London newspaper *The Independent*, 24 May 2004)

Yet, to today, many people are still terrified of radiation. Partly because it is invisible, making it seem all the more threatening, and partly because of its potential to cause cancer and genetic deformities. Many other cancer-causing agents such as food or smoke seem innocuous by comparison. Nuclear accidents, such as the partial meltdown at Three Mile Island in the US, in 1979, or at Chernobyl, in 1986, when the reactor exploded, or at Fukushima in Japan in 2011 when a tidal wave knocked out several reactors' cooling systems causing more explosions and a slow meltdown... *of course* provoke scary headlines throughout the media. (Even without popular dramatizations of nuclear risks, such as the film *The China Syndrome*, released, by a happy coincidence (at least for the film-makers), just twelve days before Three Mile Island, in which a sinister nuclear cabal covers up evidence of radiation leaks, nuclear reactors provoke a certain kind of deeply-rooted suspicion about technology.)

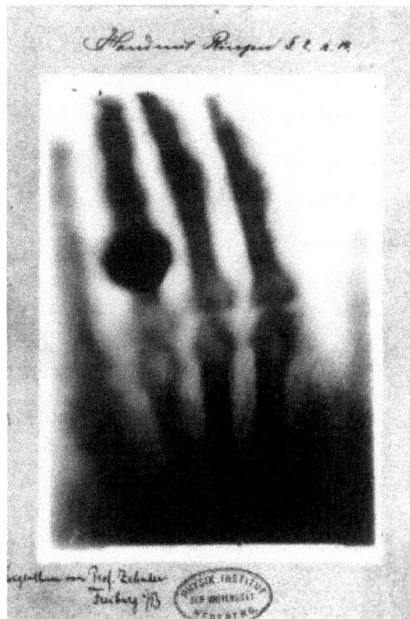

Figure 11. Hand mit Ringen (Hand with Rings). Perhaps the first public glimpse into the mysterious world of radiation was this X-ray of the left hand of Anna Bertha Ludwig (the lump is her wedding ring), taken by her husband, Wilhelm Röntgen. The framed print from the X-ray was presented to the Physik Institut, University of Freiburg, on 1 January 1896.

Nowhere is the division between popular superstition and expert reassurance more clear than in the case of nuclear science. But the first risk factor to assess with nuclear power is not 'how dangerous is it if radiation leaks out?', but the much more commonsense question of:

'Could there be an accident?' As to that, the public is regularly assured that there are so many safety systems with so many layers of security that an accident is a million million to one. Curiously though, they seem to be almost boringly frequent! Here are just some of the more famous ones.

12 December 1952. The removal of four control rods at an experimental nuclear power reactor at Chalk River, Canada, near Ottawa, led to a partial meltdown of the reactor's uranium fuel core. This is the first known major malfunction of a nuclear reactor. A million gallons of radioactive water accumulated inside, but there were no accident-related injuries.

7 October 1957. By the time the fire in Pile No. 1 at Windscale, the plutonium-production plant north of Liverpool in the UK, is put out with water, radioactive material had contaminated some two hundred square miles of the surrounding countryside. Officials banned the sale of milk from cows grazing in the area for more than a month. The government later accepted that 'at least 33' cancer deaths could be attributed to the effects of the accident.

3 January 1961. A small problem during the removal of the control rods at the core of a military-experimental reactor near Idaho Falls in the US quickly led to an explosive build-up of steam. Three servicemen were killed, one of them grotesquely impaled on a control rod.

22 March 1975. A worker using a lighted candle to check for air leaks at Browns Ferry reactor near Decatur, Alabama, touched off a fire that damaged electrical cables connected to safety systems and allowed the reactor's cooling water to drop to dangerous levels.

28 March 1979. Potentially the most serious US nuclear mishap takes place at Three Mile Island. Loss of coolant caused radioactive fuel to overheat, leading to a partial meltdown and the release of a still-disputed quantity of radioactive material.

8 March 1981. A problem-ridden nuclear power station in Tsuruga, Japan, is found to be leaking radioactive waste water for several hours. Nearly 300 workers dispatched to mop it up were heavily irradiated. The problem was only disclosed publicly six weeks after, when radioactivity was detected in a nearby bay.

4 January 1986. An improperly heated, overfilled container of nuclear material at the Kerr-McGee Corp. uranium-processing plant in Gore, Oklahoma, in the US bursts, killing one worker and allowing some radiation to leak out of the plant, resulting in more than 100 residents needing hospital treatment.

April 1986, Chernobyl, Ukraine. During a routine test, the reactor core overheats causing the building to explode and spreading a vast cloud of radioactive debris over much of Western Europe.

After that, nuclear power went rather out of fashion in Europe — although it continued to find supporters elsewhere. (Indeed, several Eastern European countries discreetly signed up for more 'bargain basement' Russian reactors in the closing years of the century, and more followed as part of the Russian 'energy wars' at the start of the new millennium.)

Truly, Chernobyl is the front-line between popular opinion, which says it was an unparalleled disaster, and expert opinion that says it was a mere hiccough in energy production. The World Health Organization commissioned a Panel of Experts to look into the health consequences, and they concluded that they were much-exaggerated and trivial.

Chernobyl seems, like many other nuclear power 'near misses', to have been an accident ascribable purely to human error plus a little institutional negligence, but the following core meltdown had a special significance because the reactor was very large. A special significance and an especially large radiation cloud. Even as 'suicide' helicopter pilots valiantly managed to put out the fire in the reactor core, and then entomb the broken reactor in a coffin of sand and concrete, a cloud of invisible radiation quickly covered most of Western Europe, stretching from the beaches of Southern Spain to the highlands of Scotland.

Tracking radiation is itself a tricky task. Although after a nuclear incident a circle will be drawn on the map around the affected plant, ten, twenty, or perhaps even thirty kilometres distant, and although large numbers of people within that circle will be forcibly moved — a quarter of a million residents in the case of Chernobyl — the actual radiation follows no such pattern. It rather follows the winds and the tides, shooting off in a concentrated plume for hundreds or even thousands of kilometres before deciding to settle. Of course no one tells you that. After all, no one has any solutions!

Once the particles land, perhaps brought down by rainfall, their journey is by no means ended. They continue to travel along food chains and biochemical pathways for years to come. Three days after Chernobyl, however, the Soviet authorities had made no announcements whatsoever. Accidents at nuclear plants, after all, were secret. Famers continued to distribute milk and other products that now contained within them the seeds of death. The only reason the Soviets eventually did acknowledge there was a radiation problem was that sensors at the Forsmark nuclear plant in Sweden, some 1200 kilometres (750 miles) away, had shot off their scales, so to speak. Not all plants are equally affected. Those with bigger leaves like lettuces, spinach, and grass generally absorb most, while crops such as potatoes, rice, and corn absorb less. Fruits such as apples and oranges also collect less radiation. Clay soils tend to trap radiation in the soil whereas sandy soils allow it to pass more quickly into plants.

After Chernobyl, the exclusion zone created (30 km) became a massive laboratory for measuring radiation effects. The trees eventually trap much of the radioactivity, as do mushrooms, some of which are said to be 1000 times more radioactive than is either normal or safe. The swallows, birds which migrate and hence are in the region only a part of the year, exhibit many tumours and genetic defects. The conifers seem to be particularly sensitive to the radiation, and have been replaced by silver birch. Many studies have been made, but as ever the conclusions are often contradictory and opposed. However, one theory is that generally the plants and animals that are the most ancient in origin, that is with the most simple and hence resilient genetic code in their cells, survive radiation best. An ancient line of wild horse has been successfully reintroduced, adding a touch of exoticism.

Nowadays, tourists regularly tour the 2000 square km national park that is the exclusion zone, admiring the great sarcophagus itself, and the abandoned apartments and ruined fairgrounds of the nearest towns. But if they are offered sparrow and mushroom pie with side servings of locally produced spinach, they should decline and ask for apple pie instead. And in any case, by 8pm they must have left the zone through military style checkpoints, leaving the wolves and wild boar to roam the ruined streets in peace.

So... how dangerous was that radiation anyway? The nuclear industry, an industry with research funds stretching into tens of billions of dollars, not to mention its helpful risk management experts, has its own view which it promotes vigorously. Because the answer to that question is crucial to the economic viability of nuclear power. Not least because even when nuclear reactors don't blow up, leak, or melt down, they create massive quantities of radiation.

> 'There is no way to put it on earth that's safe. As it leaks into the water over time, it will bioconcentrate in the food chains, in the breast milk, in the foetuses, that are thousands of times more radiosensitive than adults. One x-ray to the pregnant abdomen doubles the incidence of leukaemia in the child. And over time, nuclear waste will induce epidemics of cancer, leukaemia and genetic disease, and random compulsory genetic engineering. And we're not the only species with genes, of course. It's plants and animals. So, this is an absolute catastrophe, the likes of which the world has never seen before.'

That's how Helen Caldicott, an undisguised opponent of the industry, puts it anyway. Indisputably, the cost of dealing with all the radioactive waste it generates is the ball and chain on the industry. In fact, it is only able to survive at all by giving an optimistic assessment of the dangers of low-level radiation. Farmers and fishing communities who find a nuclear plant springing up in their midst, pumping slightly irradiated water into the seas or rivers, and releasing slightly irradiated steam into the atmosphere, have to be reassured that the amounts are inconsequential compared to 'normal' background radiation.

Secondly, nuclear reactors employ perhaps surprisingly large numbers of people and it takes money to persuade someone to work in a hazardous environment. It would be so much easier, read profitable, to run a nuclear reactor if radiation was really not that hazardous after all! For instance, it would lower staff costs. But maybe not very much, because it turns out that if our collective image of a nuclear power station includes highly trained technicians in white coats, the reality is of teams of casual workers being recruited off building sites and farms for short contract work. Take, for example, the world's third largest generator of nuclear power, the apparently high-tech Japanese. According to Japan's Nuclear and Industrial Safety Agency, Japan's nuclear plants were staffed almost entirely by casual workers! In March 2010, of the approximately 83,000 workers in the 18 commercial nuclear installations, 88% were contract workers. At the Fukushima complex, the proportion of casual staff was even higher.

The people who really keep nuclear power plants running are not the physics graduates in the high-tech control rooms, but people paid by the hour, who otherwise might work in agriculture, on building sites, or on roads. They do jobs like cleaning radioactive dirt off the sides of reactors' dry wells or mopping out spent-fuel ponds with mops and rags—all the while factoring higher rates of pay (in Japan wages start at $350 a day and may go up to $1000 for just two hours) against perceived dangers in the work. While there is an elite of nuclear industry employees, highly trained and indeed highly monitored for radiation, these workers are non-unionized and frequently obliged to falsify their radiation records.

In Japan, Tetsuen Nakajima, the chief priest of a 1,200-year-old Buddhist temple, once attempted to set up a union for this huge pool of nuclear workers. Amongst the unions strikingly modest ambitions were that workers should not be made to lie to government inspectors about safety procedures, and that workers' radiation records should not be falsified to enable them to work longer. But his embryonic union was soon dissolved when unknown thugs appeared, threatening violence again anyone who joined. As Tetsuen Nakajima put it in comments to a newspaper, 'They were not allowed to speak up. Once you enter a nuclear plant, everything is a secret.'

So, is there a conspiracy amongst nuclear scientists to hide the fact that radiation is actually dangerous? Not really. The first thing, as every nuclear physicist will tell you, is that radiation is entirely natural. Radiation is all around us, in the rocks, in the sunlight, in the soil, why there's even a bit in bananas! The second thing they will earnestly explain is that our friendly Sun produces huge quantities of radiation because, after all, it is ALSO a nuclear reactor, busily fusing atoms together with heat energy as a by-product. (Although they probably won't mention that the natural radiation from the Sun would still kill all life on Earth were it not for the happy coincidence that Earth has a large molten iron core, which generates a protective magnetic field around the planet shielding us from the radiation.)

Another reassuring thing they will say is that although, yes, one spooky fact about radiation is that some of it can go through brick walls and even serious metal cladding, most of it is feeble stuff. A piece of paper, or more to the point the human skin, is enough to block it. The dangerous radiation produced by nuclear plants will be locked away securely in steel drums deep underground.

All this is true. And so the final thing nuclear experts say is that fear of nuclear power is out of all proportion to the actual risks, born of ignorance and superstition. Back in the 1980s, Some 250,000 workers (including both army conscripts and volunteers) who took part in clean-up operations around the Chernobyl reactor were exposed to high levels of radiation, with another 350,000 or so exposed to lower levels later. Five million people were exposed to radiation in Belarus, Ukraine, and Russia.

And so, following Chernobyl, the UN organized a Panel of Experts (literally, that's what they call it) to settle the matter. Did one million people die after the reactor exploded – or not? It certainly sounds like an important question to try to get the answer to right.

The United Nations Scientific Committee on the Effects of Atomic Radiation itself devoted considerable time if not energy to answering them. Its report surprised many people when it came out. First of all, it found that the effects of the meltdown and explosion of the nuclear reactor at Chernobyl, events themselves described as one-in-a-million-year risks, were not nearly as bad as people had feared.

'There is no evidence of a major public health impact attributable to radiation exposure 20 years after the accident, and no evidence of any increase in cancer or leukaemia among exposed populations.' (UNSCEAR, The Chernobyl Accident: UNSCEAR's assessments of the radiation effects)

The World Health Organization too concluded that while a few thousand deaths may be caused over the next 70 years by Chernobyl's radioactive cloud, this number 'will be indiscernible from the background of overall deaths in the large population group' (World Health Organization, 'Health Effects of the Chernobyl Accident and Special Health Care Programmes', 2006). Who, pardon the pun, is one to believe? Could the World Health Organization itself be part of a nuclear conspiracy? The *Guardian* newspaper's resident Energy and the Environment expert, George Monbiot, mocked claims by anti-nuclear campaigners, such as Helen Caldicott, that the WHO was biased:

> 'Now, on these questions that Helen raises, I mean, if she's honestly saying that the World Health Organisation is now part of the conspiracy and the cover-up, as well, then the mind boggles… If them and the U.N. Scientific Committee and the IAEA and – I mean, who else is involved in this conspiracy? We need to know.'

We can all laugh at that. But Monbiot should have reminded his readers that there is a well-known link between the World Health Organization and the nuclear lobby. There is the founding agreement between the WHO and the IAEA, 28 May 1959 at the 12th World Health Assembly, clause No. 12.40 which says plainly:

> '[W]henever either organization proposes to initiate a programme or activity on a subject in which the other organization has or may have a substantial interest, the first party shall consult the other with a view to adjusting the matter by mutual agreement…'

The World Health Organization may be interested in promoting the world's health (although it does not seem to be very good at it) but certainly here there is a problem if it has to also reconcile its reports with the official IAEA's purpose which is: 'to accelerate and enlarge the contribution of atomic energy to peace, health and prosperity through-out the world.' This kind of teamwork meant that its investigation was headed by scientists who had already made their names denying that the atomic bombs dropped on Japan caused the health effects that everyone else thought they had done. That was good news of course for both the Japanese and American governments, and indeed for the Japanese people themselves – if true. Because, clearly, there are plenty of serious people who say that radiation is deadly stuff all right.

Twenty years after the nuclear industry's worst accident at Chernobyl, Evgenia Stepanova, of the Ukrainian government's Scientific Centre for Radiation Medicine, says: 'We're overwhelmed by thyroid cancers, leukaemias and genetic mutations that are not recorded in the WHO data and which were practically unknown 20 years ago.' In one region of Ukraine, the Rivn, 310 miles west of Chernobyl, doctors say they are coming across an unusual rate of cancers and mutations. 'In the 30 hospitals of our region we find that up to 30% of people who were in highly radiated areas have physical disorders, including heart and blood diseases, cancers and respiratory diseases. Nearly one in three of all the newborn babies have deformities, mostly internal', said Alexander Vewremchuk, of the Special Hospital for the Radiological Protection of the Population in Vilne. At the children's cancer hospital in Minsk, Belarus, and at the Vilne hospital for radiological protection in the east of Ukraine, the specialist doctors insisted that they were seeing highly unusual rates of cancers, mutations, and blood diseases, and they were all linked to the Chernobyl nuclear accident years earlier.

And then came the earthquake and tsunami in Japan, followed by 'problems' at the Fukushima plant with cooling, not to mention the partial meltdown of several reactors—Chernobyl all over again! And not even one million years passed...! With Tokyo's water testing radioactive and half a million people forcibly evacuated from their homes, the nuclear lobby had to re-explain the 'real risks' to the public once again.

Writing on one of the more influential English language websites (because free) shortly afterwards, the London *Guardian*, one independent expert, Melanie Windridge, explained, alongside an editorial resolutely defending the industry, that, 'despite Fukushima', nuclear power remained one of the safest and cleanest ways to generate power. Indeed, said Dr Windridge, who is typical of nuclear experts but not otherwise remarkable, there had been problems at the Fukushima plant with cooling, gas explosions (not nuclear), and radiation leaks—all serious issues, but so far no one has died. Not one! The immediately preceding earthquake and tsunami, by comparison, killed more than 10,000 people, she added, not quite managing to avoid sounding pleased.

But by now Dr Windridge's mantle of academic independence begins to wear thin. Her byline for the article offered only that she was freelance science communicator and academic visitor in nuclear fusion research at Imperial College London, a university as we have seen with a particular speciality of downplaying environmental hazards, not to say with links to the British nuclear industry. Downplayed too is the fact that she actually works as a researcher at the Culham Science Centre in Oxfordshire in the United Kingdom, for an organization called the Centre for Fusion Energy... owned and operated by the United Kingdom Atomic Energy Authority.

Instead of talking about her own nuclear links, she explains that the Fukushima disaster actually shows, contrary to unscientific folk, how safe nuclear reactors are. 'Reactors designed half a century ago survived an earthquake many times stronger than they were designed to withstand, immediately going into shut-down (bringing driven nuclear reactions to a halt).'

'I do not wish to trivialise the problems at Fukushima. I dislike the radioactive waste and safety issues of nuclear fission as much as anyone, which is why I work in research into a new form of nuclear energy —fusion. It's the energy source that powers the sun and has none of the downsides of fission. Fusion will produce abundant energy cleanly and safely, but it is not yet ready. With continued political and financial support we hope to have fusion power stations by the 2050s.'

Send your money in now, folks! But then back to the elephant-in-the-room question. How dangerous is all that nuclear radiation? Dr Windridge allows that leaks are potentially serious. But then it is back to reassurances—remember that we are subjected to background radiation every day as a result of natural processes. There's radiation from granite rocks, there's radiation from cosmic rays affecting anyone regularly flying high-altitude routes. 'And then people routinely and willingly expose themselves to large amounts of radiation for medical checks, with dental x-rays providing perhaps the highest doses, often for purely cosmetic reasons', Windridge adds. (Implying that household electricity is always used for extremely important things...)

By comparison, nuclear facilities are famously stringent and releases of radiation are taken 'extremely seriously'. Ironically, however, exactly these precautionary limits can cause unnecessary alarm amongst the hoi polloi masses. For example, says Dr Windridge, take that scare about contamination of water supplies after the 'problems' at the Fukushima reactor. There were recommendations for restrictions on drinking water, yet 'the radiation dose received by someone drinking Tokyo water would have been less than that from moving to Cornwall and living there for a year'!

But we should note this point in passing. The key difference in radiation is whether it is external to the body, as in background radiation produced by rocks, or internal—as when you drink water or eat contaminated food. A second fact about radiation that nuclear experts seem not to know is that how dangerous radiation is critically depends on whether it is experienced all at once or spread out over a long time. After all, as they say, life on Earth has evolved to tolerate a certain amount of radiation. The body has defence mechanisms that react to deformations of the genetic code in cells. But here, one dose spread out over the 500,000-odd minutes of the year is cheerfully compared to the same dose absorbed in the minute it takes to drink a glass of water! Windridge may not know the difference, but the body certainly does.

If you inhale an invisibly small particle of plutonium, the surrounding cells receive a very, very high dose of radiation. Most of the cells near the plutonium die, because it's an alpha emitter. However, some cells on the periphery remain viable. They mutate, and the DNA, the regulatory genes, are damaged. Inevitably, although perhaps years later, that person develops cancer, although the body may yet fight off the cancerous cells. The same is true for particles of caesium-137, which heads for the brain and muscles; or strontium-90 that goes to bone, causing bone cancer and leukaemia, or radioactive iodine that goes to the thyroid.

In the weeks following Fukushima, for example, the levels of caesium-137 in the village of Iitate, some 40 kilometres northwest of the plant, were measured at more than twice the levels that prompted the Soviet Union to evacuate people near Chernobyl. Radioactive iodine was found in the tap water in all of Tokyo's 23 wards. Monitors detected tiny radioactive particles from the reactor site that had spread across the Pacific to North America, the Atlantic, and even Europe.

The *New Scientist* was very concerned. An article in the relatively specialist magazine just a few weeks after the plant started malfunctioning calculated that the relative radiation releases of Chernobyl (10 days duration) and Fukushima (the calculation being made 24 days after the incident) indicated that Fukushima had already produced more of the isotopes largely responsible for approaching one million deaths due to Chernobyl, not to mention countless illnesses and transgenerational mutations than the earlier disaster. (To be precise, about 1.7 times the amount of iodine-131 and 1.4 times the amount of caesium-137).

And so, 'each month that passes will add two more Chernobyls of Iodine-131 and Caesium-137 to our environment, by air alone, if releases remain at current levels. The hazards of course will catastrophically increase if the plant becomes so radioactive that workers can no longer visit the site to keep what's left of the reactors cool.' The *New Scientist* added ominously that these were just the figures for atmospheric pollution — multiply them when you add in the radioactive water being pumped into the Pacific every day.

'And these releases are expected to continue for months, if not years. When are the "authorities" and mass-media going to stop lying to us and belittling the extreme seriousness of what is happening at the Fukushima plant?', asked the *New Scientist*. That's one view. However, for other nuclear experts, this spreading out of radiation across continents is a good thing. Melanie Windridge, for instance, offers that, with radiation, anything that flows into the ocean 'either by accident or to relieve storage problems on land' will be greatly diluted. But what does she mean? That only if you drink the whole ocean will you get a dangerous dose? Or if you eat a fish that contains a grain of tiny but highly radioactive material it will have been rendered harmless by virtue of having spent some time in the sea? Because, of course, radioactive compounds are not diluted in the oceans but merely spread around. Although, it's true, it makes it hard to tell what caused who to fall ill and die.

Estimates of radiation deaths are all about statistical probabilities. Yet if it is at bottom a vague science of statistical generalizations, some generalizations are more convincing that others.

> 'The WHO Expert Group placed particular emphasis on scientific quality, using information mainly in peer-reviewed journals, so that valid conclusions could be drawn.'

Here's what it calls 'valid' reasoning: it considered that most of the people described as dying from radiation had really died from 'overuse of alcohol and tobacco, and reduced health care'. The methodology is that advanced by those who warn members of the general public unused to scientific method to remember that, if you find someone apparently dying from something rare and strange like nuclear radiation, you should first of all check their cupboard for chocolate biscuits. It could be that they are really dying from those. Not to forget that even those really dying from nuclear radiation might have died from cancer 'anyway'. 'The Expert Group concluded that there may be up to 4000 additional cancer deaths among the three highest exposed groups over their lifetime… 3–4% above the normal incidence of cancers from all causes.' In sum, the WHO/IAEA estimate that the Chernobyl nuclear accident has killed 58 people and will lead, in the years to come, to 'at most' an additional 4,000 fatal cases of cancer.

The WHO uses a statistical method that dismisses any health effect from radiation below a certain level and, among those it considers did receive significant dose, it predicts precisely 6,158 additional cancers in 50 years which, among the two and a half million cancer cases expected normally in that population over half a century, would be invisible. This is because there's always a causality problem when looking at certain deaths that can have multiple causes (was cancer caused by radiation, smoking, or bananas?).

> 'Given the low radiation doses received by most people exposed to the Chernobyl accident, no effects on fertility, numbers of stillbirths, adverse pregnancy outcomes or delivery complications have been demonstrated nor are there expected to be any. A modest but steady increase in reported congenital malformations in both contaminated and uncontaminated areas of Belarus appears related to improved reporting and not to radiation exposure.'

If cancer deaths have increased by more than that, they say (which they have), then it is due to people smoking and drinking more, eating unhealthy foods (like chocolate biscuits) and... wait for it... worrying unnecessarily about radiation. 'Radio-phobia' not radiation is the problem. For the WHO experts only one cancer seems to be directly and unambiguously attributable to the Chernobyl cloud — thyroid cancer in children. This is because there is no other cause for the disease than nuclear radiation. But even here the figures are not as bad as they seem. Rather ingeniously, the experts argue that almost all of the cases are not real 'new cases' but merely due to increased reporting of the illness. (You could say that about all statistics, but you won't have many left at the end of the day.) In previous years many children apparently had cancer of the thyroid but no one noticed. In fact, the WHO concludes pompously, the nuclear accident caused only nine children to die from the disease, and those deaths were really the fault of the then communist authorities for not banning milk sales straight away.

Otherwise only people who had received high doses, such as those involved in the clean-up operations immediately afterwards, could really be said to have radiation induced illnesses, the WHO experts reported. As to these, the United Nations Scientific Committee on the Effects of Atomic Radiation assessed in a report in 2000 that just 134 liquidators received radiation doses high enough to be diagnosed with acute radiation sickness (ARS). And, of them, just 28 persons died in 1986 due to radiation poisoning. 'Other liquidators have since died but their deaths could not necessarily be attributed to radiation exposure.'

As for claims, indeed orphanages in Ukraine and Belarus, full of children with tragic and often grotesque birth deformities, the WHO is emphatic. Chernobyl caused no 'congenital malformation' or birth defects in babies at all. 'Given the low doses' of radiation the parents would have encountered, they could not have done. QED. But... but... the numbers of such children have shot up! Ah yes, but they have shot up all over Ukraine and Belarus. And some regions received more radiation than others. Therefore the cause for the increase must have been something else. Maybe increased reporting, radio-phobia... and chocolate biscuits.

So, if that's the expert assessment, why are there massively different figures — a million dead, orphanages full of children with horrible birth defects — still floating around? Eventually it comes down to people actually dying. As the police say, if you've got a dead body, you've got a crime to investigate. With Chernobyl, it seems we have plenty of bodies now. 'At least 500,000 people — perhaps more — have already died out of the 2 million people who were officially classed as victims of Chernobyl in Ukraine.' Or so said Nikolai Omelyanets, deputy head of the National Commission for Radiation Protection in Ukraine. Adding that infant mortality alone increased 20% to 30% because of chronic exposure to radiation after the accident.

Such figures appear to make a nonsense of the optimistic report of the World Health Organization on the effects of the radiation cloud after Chernobyl. Or they might do if the WHO included such figures in its reports, but it doesn't. The UN has its rules, and reports have to be in certain journals and in English or they will not be accepted for consideration. If a nuclear plant explodes in the United States they may get around to assessing the consequences more thoroughly.

Nikolai Omelyanets says his information has been ignored by the IAEA and WHO. 'We sent it to them in March last year and again in June. They've not said why they haven't accepted it.' An IAEA spokesman said he was confident the figures in the WHO report were correct. 'We have a wide scientific consensus of 100 leading scientists. When we see or hear of very high mortalities we can only lean back and question the legitimacy of the figures. Do they have qualified people? Are they responsible? If they have data that they think are excluded then they should send it.'

The WHO is not always so cautious with its statistics, however. As described in Chapter 4, facing the problem of colds and flu, where the multi-billion dollar industries of vaccine production are involved, here, its response to unclear causation and small probabilities is rather different. Take the SARS epidemic for example. Here, the World Health Organization issued an emergency warning (despite it being the weekend) declaring the sickness 'a worldwide health threat', after four deaths related to pneumonia and 'another five in an outbreak of a similar infection in a province of China', although the two had 'not yet been definitively' linked. And, by comparison, the unremarkable flu of 1999 killed 21,000 people in Britain alone that winter.

Where exploding nuclear reactors produced advice to drink less and be positive, cold germs produced dire warnings. That newly appointed UN Co-ordinator for Avian Flu earnestly hoped measures could keep the death toll to between 5 million and 150 million. Large transnational pharmaceutical companies were asked to start preparing 'pre-pandemic' vaccines. In the US alone, $7 billion were allocated to help pay them. Yet a year later the worldwide toll from the disease was put at… just 262. A million people there is 'statistically insignificant…', 262 people here is a worldwide alert. Or could it just maybe be 'bad science'?

The nuclear industry and its regulators calculate risk on the basis of the likelihood of an accident for any one operating year. In the case of the design of the first four reactors at Fukushima, the Japanese Nuclear

Energy Safety Organization estimated in 2002 the 'frequency of occurrence of a core damage accident is 1/100,000 or less per one year for one reactor and the frequency of occurrence of an accident leading to containment damage is 1/1,000,000 or less per one year for one reactor'.

Somehow, however, in 2011 it got four reactors in that position, odds that must have been one million million million million against! Vanishingly small indeed. Funny how it happened, isn't it? Or take the risk of a major earthquake causing a nuclear plant to meltdown. In Japan the risk of magnitude 9.0 earthquake is a circa 100-year event. That's clearly something that could happen at any moment. But in Japan, the plants were required only to withstand tremors one thousand times less powerful. At Fukushima, in particular, a sea wall was built to cope with seven-metre waves. But the chance of greater waves than that was calculated as being highly likely. And given that only a few decades, rather than millennia, separate the accidents at Fukushima, Chernobyl, and Three Mile Island (which were also enthusiastically described as being at minimal risk of core damage), it is clear that nuclear operators and/or regulators are significantly underestimating the inherent risks associated with nuclear technology. Just as they systematically understate the risk from radiation, and for much the same brass-necked reason.

But here is what one former UK government minister who once backed nuclear power now says: 'At no stage, as a minister, could I rely on being told the truth either by the Industry itself, or by my own civil servants who may or may not have known it themselves' — Tony Benn, who as a minister in the 1960s was in charge of nuclear power.

Disinformation on the dangers of nuclear radiation is as old as the discovery of the phenomenon itself. Even after the first nuclear bomb had been dropped on Hiroshima, a General Groves was still prepared to assure the United States Congress that a team of scientists had found no traces whatsoever of radiation in Hiroshima and that, in any case, radiation poisoning was 'a very pleasant way to die'. Unfortunately, a month later, an Australian journalist smuggled a report out — past the US censors — from Hiroshima. This gave a different picture of the day...

> '...patients just wasted away and died. Then people... not even here when the bomb exploded, fell sick and died. For no apparent reason their health began to fail. They lost their appetite, head hair began to fall out, bluish spots appeared on their bodies, and bleeding started from the nose, mouth and eyes. We started giving vitamin injections, but the flesh rotted away from the puncture caused by the needle.'

It certainly didn't sound very pleasant... and in every case the patient died.

Chapter Twelve

African Art or High Street Kitsch?

A cautionary tale (or two) of what happens when too much objectivity is assumed in matters best left at least a little subjective

Subject: Fine Arts

What we're supposed to think:

'Aesthetic theories must prove themselves valid more in the sense that scientific theories must, than in the sense that ontological postulations can. Aesthetics is the science of art…'
—Professors Friedrich Ulfers and Mark-Daniel Cohen, 'The Effect of Nietzsche's Aesthetics on the Art of the Twentieth Century'

What you're not to say:

'To feel beauty is a better thing than to understand how we come to feel it.'
—George Santayana, *The Sense of Beauty*

What do you do if, one day, rummaging through some long-forgotten trunk in the attic, you come across a box of strangely carved wooden statues, some with scary eyes and others in strange contortions? You call in an art expert of course, because these statues could be art. More to the point, they might be valuable. On the other hand, they might just be junk, stuffed there years ago by someone who had just sold another fifty identical ones in the flea market. Unfortunately, it would be hard for a non-expert to tell the difference. Yet in money terms there is a huge difference between art and mere woodwork, between abstract masterpiece and an enthusiastic effort done with a toothbrush. But don't be too quick to assume that the difference is easy to pinpoint, one of subject matter, or attitude, let alone skill. Because, in the best tradition of science, aesthetics does not follow any rules.

Take the case of ethnic art, like African masks or sculptures. These by their very nature can be both crudely made and crude in appearance — because they are linked to sexual rites and traditions in simple societies with no role of elaborate and delicate works. As such, they are easily and cheaply mass-produced for tourists too — maybe in a garage in the neighbouring town. So how do you tell which is a work of art and which is a work of complete cynicism? That's when you need an expert.

Someone like Bernard Dulon, who pays good money for such works, and has his own ethnic art gallery in Paris. Bernard has said that he can tell 'straight off it's the real thing' — because every work of ethnic art has its own 'aesthetic, emotional and ethnological language'. Or, take another contemporary French art dealer, Renaud Vanuxem, who says he can always tell a fake from a genuine artefact because 'although it might look right, it doesn't feel right, it doesn't have soul.'

This is pretty metaphysical stuff. Lots of people no longer think human beings have 'souls' — let alone that pictures or African masks have. Yet now philosophers of art want us to accept that works of art have 'soul' and forgeries (or just lesser works — 'bad art', like I might do) don't. But can the experts justify such distinctions to the rest of us? And there are some rather grand theories which say that they can. 'Baumgarten's science of aesthetics', for instance, which insists that judgments about beauty and good taste are 'rational', that they really can be 'right or wrong', and that certainly not every view is equal. That is why is Marcel Duchamp's toilet bowl is considered art and that's why Michelangelo was wrong when, according to Vasari, he took a totally uninspired marble cupid (the one that he had just carved), buried it for a time to give it an antique look, and sold it as an ancient sculpture. Because that was the act of a forger, not a great sculptor.

Confused? Denis Dutton, Professor of Philosophy at the University of Canterbury in New Zealand, sometime editor of the well-known website *Arts and Letters Daily,* and author of a book called *The Art Instinct,* once tried to explain the difference with regard to one of the art world's most famous scandals—how for decades experts praised the works of the twentieth-century forger Hans van Meegeren, who made increasingly implausible 'copies' of a much valued seventeenth-century Dutch painter, Vermeer. The curious thing about these 'fakes', cheats, copies, call them what you will, is that they were not copies of actual paintings but were only forgeries in that Hans van Meegeren passed off his work as being previously unknown 'rediscovered' works of the Old Master. (Another curious thing was that when Hans van Meegeren was put on trial after the Second World War for having sold national treasures to the Nazis, he was able to prove his innocence by painting 'another masterpiece' in his jail cell.)

As Professor Dutton puts it (and he would know) the 'fakes' all had their own style, and didn't actually look like the rest of the 'known' works of Vermeer. But because Hans' first effort had become part of the supposed catalogue of Vermeer, all his later works seemed to match too. Of course they did! In fact, the fakes looked more like twentieth-century German expressionist paintings than they ever looked like seventeenth-century Vermeers. So the fraud was not in lack of style or lack of originality, but simply in using a fake name. It seems indeed that his forgeries did have 'soul' but maybe not the one expected.

Well, Hans van M. shouldn't have borrowed a famous person's name. That's why his art is a definite 'fake'. And for economists and art dealers, that is enough of an answer to what is 'real art'. But what about if we go back to the case of 'Tribal Art'? Here there are no famous names, only mysterious objects and unknown traditional meanings. Indeed, some art experts think that the key to separating 'real' tribal art from the stuff made in a garage down the road for tourists is their function. Specifically, that the real things are not made with any interest in 'us', the Westerners who later come across it, but rather serves a special, esoteric purpose—that is a purpose known only to those in the special circle of the tradition. In short, as Denis Dutton has put it, the very finest works of tribal art, New Guinea art, or African art 'have a lack of any interest in our perception at all, they're sort of in another world'.

Nonetheless, African witchdoctor masks or whatever are also said to be art because they do express something important for their makers and the 'original audience', with shared values. Like Western paintings, they are supposed to also be the expression of something unique by one human being.

Art experts today rail against what they sometimes call 'kitsch' — art which borrows the superficial suspects of great art, adds expensive frames, and conveys 'simplistic' messages. The sort of paintings and statues ordinary people actually like. Surely here we can agree that this sort of stuff is never going to be 'art'? Yet, equally, most unfortunately for collectors, many of the great artists (like Michelangelo) have been reduced to operating in rather 'inauthentic' ways — burying statues to make them look old, or dashing off pictures and sketches for landlords and rich benefactors just to keep the debts at bay. In fact, most of the 'great works of art' were paid commissions — for the Church or for the local landowner. Many 'Great Paintings' started off very plainly as portraits commissioned by rich aristocrats of themselves to record their success in life. However, this money-minded origin has not stood in the way of these works being accepted. Only recently have artists enjoyed the luxury of being just able to 'create'!

Literally, to be authentic is to be real, and not fake. Yet even when an art object is not a fake, it can still be rather inauthentic. So the existentialist view is often linked to the notion used by philosophers and artists in debates on aesthetics. These philosophers seem to use the word to mean something like imaginatively 'original'. Existentialists — Jean-Paul Sartre and Edmund Husserl — put a priority on 'authenticity' but it's not very clear what this adds up to beyond the slogan. To make matters worse (more complicated), Sartre himself insists that 'the real is never beautiful' before adding: 'Beauty is a value applicable only to the imaginary... that is why it is stupid to confuse the moral with aesthetic.'

At least making barefacedly false claims about your 'art' is more straightforward. Fake art is as old as the real stuff, but art forgery only really came into its own when the large-scale collection of antiques by wealthy individuals and municipal institutions developed, along with a rising cult of artistic 'personalities', such as Leonardo da Vinci, Van Gogh, or Picasso.

Particularly during the twentieth century, the ever-increasing emphasis on the financial value of works of art encouraged many master forgers. One such was Alceo Dossena of Cremona, a sculptor whose speciality was to use the carving techniques not of his own time, but those of antiquity, the Middle Ages, and the Renaissance. His work is acknowledged to be of the first rank and the highest quality and not mere imitation of the styles he admired; rather he was inspired by them to the creation of his own, similar works. One of his paintings, the Virgin and Child, painted in the fifteenth-century Florentine manner, is still in the distinguished Victoria and Albert Museum in London. Yet, fine though they are, they are also all forgeries.

A bronze horse, purportedly an antique Greek work, and an Etrus-
can warrior are two famous cases of forged sculpture brought to light
at New York's Metropolitan Museum of Art. For years they had fitted
in very well there! And surely thousands of lesser faked objects are
displayed in private and public collections. Perhaps surprisingly,
despite modern technological advances, much forgery remains
impervious to detection by other than empirical means. When two
sketches, both entitled *Tête de Jeune Fille* (Head of a Young Girl), drawn
with the same ink, and on the same kind of paper, but only one by the
great impressionist Henri Matisse, appeared on the market, the
evaluation had to be primarily aesthetic. Fortunately, in this case, as
one art expert put it, 'the copy is so inferior, so servile, that in my eyes
the spirit of trickery is blatant. The hand follows the master, but has not
the surety, nor the rhythm, the pen cannot kiss the paper with the same
gentleness.' Indeed, looking at the sketches in Figure 12, one can but
agree. But, er… which one is the original?

You might have thought that technology would come to the aid of
the collectors, but then the discovery of forgery results in a curious
phenomenon—a work of art may be considered a priceless masterpiece
one day and worthless the next. For many wealthy collectors, it is better
not to know.

Figure 12. Which one is the 'real thing'? Here are two sketches of *Tête de Jeune
Fille* (Head of a Young Girl), drawn with the same ink, and on the same kind of
paper, but only one is by the great impressionist Henri Matisse. (© Henri Matisse,
Tête de Jeune Fille, Succession H. Matisse/ archives Michel Maket.)

Instead, a vaguer and imminently more flexible expertise in the styles
and aesthetics of various periods is the principal tool of the art expert.

Concepts such as 'artistic clumsiness', a 'jumble of styles or motifs', or 'a discernible emphasis on the aesthetic values of the forger's own day' are the armaments in the weaponry of the expert authenticators, much more than mere physical tests. For example, the nineteenth-century Russian creator of the much admired engraved golden tiara of Satapharnes (in the Louvre), supposedly a Scythian head-dress of the third century, was exposed as the pusher of just so much (gold) junk because they had used motifs from a nineteenth-century publication.

However, other forgeries are often indistinguishable from the 'real thing' by these subtle factors. Rumours constantly circulate that the *Mona Lisa*, which hangs with its enigmatic smile in the Louvre, is a value-less simulacrum, hanging there behind bullet-proof glass while the real one is actually locked away in a vault, 'for safety'. Art pilgrims who pause in front of it and feel a strange sense of contact with the soul of the artist who painted it would certainly be annoyed if they found out later that it was actually done by a jobbing copyist. But the French would do such a thing to their art lovers — they are unashamed to offer spanking new repro 'cave paintings' in modern 'caves' to the masses, while the real ones are kept for an elite, the — guess who — experts, to admire.

Ironically, though, the 'real' *Mona Lisa* painting's status as the world's most valuable itself hangs but by a slender thread, as there are five other rival versions which insist that they are the original. Perhaps the best thing to do would be to have a UN convened panel of experts examine the six paintings, scientifically authenticate the best one, *and destroy the rest?*

Mad? But good sociology. Nowadays in China, there is a popular new TV show called *Collection World* in which 'amateur collectors' meet professional experts. The amateurs get their most prized Qing dynasty vases, or delicately carved wooden chests, authenticated — authoritatively dated and valued by the experts. The twist is, if the experts say the work is a modern reproduction — a fake — then the owner must immediately take a sledgehammer to their pride and joy — and smash it to pieces!

In many areas of life, we are in a similar position to the collector on the show, making irrevocable decisions on the strength of others' advice. And yet, for sure here, the 'expert panel' was constructed of people who actually had no knowledge of antiquities. How many ancient vases and works of art have been smashed for the amusement of TV audiences, we will never know. That's showbiz! If watching it makes you want to cry out, 'Stop! That's a work of art, you cynical frauds!', then the programme is twice as much fun. And, there, even the best experts make mistakes... Tune in next week!

Afterword:
Paradigm Shifts

*Popper's new paradigm, or why science needs
to be more political than it likes to pretend*

Subject: Philosophy of Science

What we're supposed to think:

'Physicists have shown that all matter consists of a few basic particles
ruled by a few basic forces. Scientists have also stitched their knowledge
into an impressive, if not terribly detailed, narrative of how we came to
be. The universe exploded into existence roughly 15 billion years ago
and is still expanding outwards. About 4.5 billion years ago, the debris
from an exploding star condensed into our solar system. Sometime
during the next few hundred million years, single celled organisms
emerged on the earth. Prodded by natural selection, these microbes
evolved into an amazingly diverse array of more complex creatures,
including *Homo sapiens*.

I believe that this map of reality that scientists have constructed, and
this narrative of creation, from the big bang through the present, is
essentially true. It will thus be as viable 100 or even 1,000 years from
now as it is today. I also believe that, given how far science has already
come, and given the limits constraining further research, science will be
hard-pressed to make any truly profound additions to the knowledge it
has already generated. Further research may yield no more great
revelations or revolutions but only incremental returns.'
—From a talk called 'Why I Think Science Is Ending' by John Horgan,
sometime editor of the *Scientific American* magazine.

What you're not to say:

'Scientific "facts" are taught at a very early age and in the very same
manner in which religious "facts" were taught only a century ago. There

is no attempt to waken the critical abilities of the pupil so that he may be able to see things in perspective. At the universities the situation is even worse, for indoctrination is here carried out in a much more systematic manner... In society at large the judgement of the scientist is received with the same reverence, as the judgement of bishops and cardinals was accepted not too long ago. The move towards "demythologization," for example, is largely motivated by the wish to avoid any clash between Christianity and scientific ideas. If such a clash occurs, then science is certainly right and Christianity wrong. Pursue this investigation further and you will see that science has now become as oppressive as the ideologies it had once to fight.'
—Paul Feyerabend, from his talk entitled 'How To Defend Society Against Science'

Who to believe? Is science really the systematic piecing together of facts about the universe — a process now almost complete — or an unending assembling and disassembling of competing ideas and theories — politics by another name?

Almost everyone believes it is the first. More than that, almost everyone is taught it is the first. Conventionally speaking, we suppose that when experiments are conducted to 'test' theories in reality, and that when the results do not accord with those anticipated, the theory is 'disproven'. Or 'falsified' as Karl Popper puts it in *The Poverty of Historicism*. Yet nothing could be further from the truth. All scientific theories, let alone 'facts', are open to challenge, on this or other ground, from this or that perspective. Stop a moment to consider what we mean by scientific facts anyway. The most conservative scientist will accept that the body of knowledge is not so much cast in stone but organically developing. It seems, as far as textbooks go, that in thirty years every component fact will have changed: the age of the universe, how it was created, how life began, how DNA works, even how Matisse did his sketches...

Science is rather about choosing theories that suit our purposes best. Paul Feyerabend again:

> 'When Copernicus introduced a new view of the universe, he did not consult scientific predecessors, he consulted a crazy Pythagorean such as Philolaos. He adopted his ideas and he maintained them in the face of all sound rules of scientific method. Mechanics and optics owe a lot to artisans, medicine to midwives and witches. And in our own day we have seen how the interference of the state can advance science: when the Chinese communists refused to be intimidated by the judgement of experts and ordered traditional medicine back into universities and hospitals there was an outcry all over the world that science would now be ruined in China. The very opposite occurred: Chinese science advanced and Western science learned from it. Wherever we look we see that great scientific advances are due to outside interference which is made to prevail in the face of the most basic and most "rational" methodological rules. The lesson is plain: there does not exist a single argument that could be used to support the exceptional role which science today plays in society.' ('How to Defend Society Against Science')

The problem is not that politicians are making bad choices from the theories that are presented. Rather, it is the idea that experts know almost everything. Because, as Popper goes on to describe:

> '...tests can be interpreted as attempts to weed out false theories — to find the weak points of a theory in order to reject it if it is falsified by the test. This view is sometimes considered paradoxical; our aim, it is said, is to establish theories, not to eliminate false ones. But just because it is

our aim to establish theories as well as we can, we must test them as severely as we can; that is, we must try to find fault with them. Only if we cannot falsify them despite our best efforts can we say they have stood up to severer tests.'

Once the theories have been tested, they can become part of the grand edifice of scientific knowledge. But Popper says we must test theories properly, under the most difficult circumstances. For if we are uncritical, he says, then 'we shall always find what we want: we shall look for, and find, confirmations, and we shall look away from and not see, whatever might be dangerous to our pet theories.'

Karl Popper (1902–1994) is counted as one of the great philosophers of science, although he was never welcomed quite into the heart of the 'academic citadel'. He was after all a 'foreigner' to both his British and American peers. He was also a social and political philosopher of considerable stature, a staunch defender of liberal democracy (and the principles of social criticism upon which it is based) as well as of the 'open society', and, above all, an implacable opponent of authoritarianism or what is nowadays sometimes called centralized, 'big-government'.

Popper described himself as a 'critical rationalist', the 'rationalists' being thinkers such as Descartes, Leibniz and, above all, Kant. The term also signified his rejection of 'classical empiricism', as was being revived and refined by the logical positivists of the so-called 'Vienna Circle' in the 1930s. Such people insisted that actual, tangible experiments had to settle the big scientific questions, and where they could not, the questions were nonsensical. Against all these, Popper argued that there are no 'theory-free', infallible observations as empiricists ask us to assume, but rather, all observation is theory-laden, and involves seeing the world through the distorting glass (and filter) of a pre-existing conceptual scheme.

Following up on the subversive philosophy of David Hume, some centuries earlier, Popper categorically rejects 'induction', that is, the 'inferring' of general laws from particular cases, the process which is the basis of scientific method. Such inferences, he says, should play no role in scientific investigation since it is logically impossible to ever secure the 'verification' of a universal statement. You cannot prove that all rocks are heavier than water as it is always possible someone discovers a new one that isn't, or even that the properties of water itself change. Since all scientific theories are like this, making universal claims for their truth, they are unverifiable.

Sounds odd? But, as the astronomer Fred Hoyle once put it:

'The procedure in all branches of physical science, whether in Newton's theory of gravitation, Maxwell's theory of electromagnetism, Einstein's theory of relativity, or the quantum theory, is at root the same. It consists of two steps. The first is to guess by some sort of inspiration a set of mathematical equations. The second step is to associate the symbols used in the equations with measurable physical quantities.' (From an essay called 'The Nature of the Universe', 1950)

That's why no number of positive confirmations at the level of experimental testing can ever really confirm a scientific theory. Yet this has not dented the scientific edifice one little bit. Although in his essay 'Physics and Reality', published in 1936, Einstein talks of 'the verification of the derived propositions (*a priori*/logical truths) by sense experiences (*a posteriori*/empirical truths)' he immediately afterwards emphasizes that scientists 'must always be ready to change these notions – that is to say, the axiomatic basis of physics – in order to do justice to perceived facts in the most perfect way logically'. Because science knows it is really built upon inspiration and intuition, not on measurements. It does not matter how much evidence you can produce to support a theory – the next case along can still destroy it. That's the lesson learnt the hard way by Bertrand Russell's unfortunate chicken, which waited eagerly for the farmers wife each morning, confident that it had enough positive confirmations of its theory that the arrival of Farmer's wives is followed by tasty grain to eat. (Because, this morning, for the chicken, it's future is to be dinner!) The sun might not rise tomorrow and the next time you open the fridge a gorilla might leap out. It seems unlikely, but something being unlikely has no power over whether or not something happens. However, Popper makes more of the negative cases than does Hume, he offers that every counter-instance is decisive: it shows the theory to be false.

Here, as one contemporary academic, Stephen Thornton, puts it, the central thrust of Popper's argument is Socratic, knowledge-seeking becomes a process of seeking out a counter-example to demolish old theories with the intention of producing in their place better ones. Of course, in a way David Hume in the eighteenth century remains more radical than Popper in the twentieth. Hume concluded that science and philosophy alike rested less upon the rock of 'logic' and method, but rather upon the shifting sands of scientific fashion and aesthetic preferences.

The modification of the Ptolemaic System, that is the ancient Greek theories of how the heavenly bodies might be on crystal spheres, is itself a 'paradigm' demonstration of how 'falsification' does not seem to take place, let alone decide the survival or otherwise of a theory. Instead, the Ancients simply increased the number of spheres each time observations showed a problem for the theory. It is here that the really rather psychological and obscure notion of the 'paradigm shift' comes in.

Paradigm shifts, being social in nature, don't have to come out of experiments in laboratories and they don't have to be by scientists. But they surely involve discoveries nonetheless. Take the shift in our perception of the Earth created by the adventure of the Apollo moon landings described in chapter six on cosmology. Here, it is not so much the 'fact' that we live on a speck in the universe as a new way to visualize the fact that counts. An image of Saturn and its rings taken by the Voyager probe captured by chance a far-away blue dot: the Earth. But it was a chance sighting of the crescent Earth 'rising' in the dark sky, glimpsed by the Apollo 8 astronauts as they came from behind the moon after the first ever lunar orbit, that changed the way a whole generation felt about their planet.

David Scott and Richard Jurek link this particular image, along with one of the 'whole Earth', to adding emotional force to the new environmental movement—epitomized by Stuart Brand's *Whole Earth Catalog*. This featured the original 'whole Earth' image on its cover, and demanded a new kind of thinking of its many readers based precisely on that new perspective: *'We are as gods and might as well get good at it.'*

In academic terms, though, the subversive and attractive notion of the paradigm shift was first released from the laboratory by Thomas Kuhn. Once out, the theory soon spread, virus-like, from the 'physical sciences' to the social sciences, and on to the arts and even business management courses. It became so popular that people began to describe Kuhn's idea of paradigm shifts as in itself a kind of paradigm shift! We should not be too in awe of Kuhn though. Feyerabend puts it with characteristic acerbity thus:

> 'Kuhn's ideas are interesting but, alas, they are much too vague to give rise to anything but lots of hot air. If you don't believe me, look at the literature. Never before has the literature on the philosophy of science been invaded by so many creeps and incompetents. Kuhn encourages people who have no idea why a stone falls to the ground to talk with assurance about scientific method. Now I have no objection to incompetence but I do object when incompetence is accompanied by boredom and self-righteousness. And this is exactly what happens. We do not get interesting false ideas, we get boring ideas or words connected with no ideas at all. Secondly, wherever one tries to make Kuhn's ideas more definite one finds that they are false.'

However, since Feyerabend equally has complete contempt for all the Ancients, Karl Popper, and, well, just about everyone really, we should not write Kuhn off for being pompous and waffly, much less because Feyerabend says so. Even the words of the 'anti-expert' have to be treated sceptically. And meaningless or not, the idea of 'paradigm shifts' in science, which is at root the idea that scientific knowledge proceeds in fits and starts, with theories fighting to the death, as it were, against each other, is a valuable antidote to the conventional picture of science as a smooth process of the accumulation and refinement of facts, facts that then accumulate into the impregnable edifices of knowledge.

It is in his 1962 book, *The Structure of Scientific Revolutions*, that Thomas Kuhn offers a way to see how this popularly imagined lofty and impregnable fortress of scientific consensus is really just a shifting façade. Demolishing the edifice does not even require a loud bang, instead working 'jobs-worth' scientists may find, at a certain point, either that the old theory has become too complicated and cumbersome to modify, and so collectively abandon it for something that now suits them better, or else a 'split' may emerge between followers of one theory and another that is eventually decided in favour of the new theory for any number of reasons, none of them particularly scientific or dramatic.

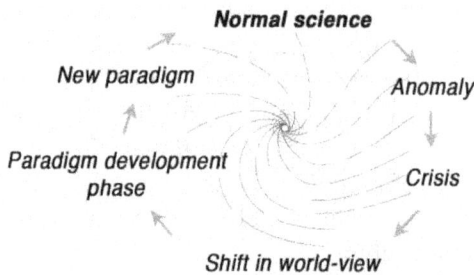

Figure 13. The relationship of paradigm shifts and normal science.

In their lifetimes, Kuhn and Feyerabend made up two viciously opposed sects, but at least they agreed that this is the true face of science. Indeed, Kuhn says, like Feyerabend, that 'a scientific community cannot practice its trade without some set of received beliefs', and that these beliefs form the foundation of the 'educational initiation that prepares and licenses the student for professional practice'. And, in *The Logic of Scientific Discovery*, Kuhn offers a convincing analogy for the state of 'normal science':

> 'Science does not rest upon solid bedrock. The bold structure of its
> theories rises, as it were, above a swamp. It is like a building erected on

piles… if we stop driving the piles deeper, it is not because we have reached firm ground. We simply stop when we are satisfied that the piles are firm enough to carry the structure, at least for the time being.'

One of the surprising lessons of the history of science is that analogies like this get us further than rigid 'rules' or theories. In area after area of science, we find the breakthroughs taking place at the level of creative analogy, not in the sifting and analysis of precise data. The equation $E = mc^2$, *energy = mass times the velocity of light squared*, is analogous to the rather mundane relationship in mechanics, *kinetic energy = mass times velocity squared* (albeit divided by two), as Douglas Hofstadter has recently argued. Hofstadter, the American 'cognitive scientist' best known for a popular book called *Gôdel, Escher, Bach: An Eternal Golden Braid* (1979), goes on to embrace 'fuzzy' thinking in many areas, including what we think of as hard science, as essential. Yet the philosophical perception of science has remained rooted in a Platonic universe of eternal and unchanging truths, a universe which the Ancients themselves recognized had only limited applicability and relevance to the world we actually live in.

So perhaps the answer to the question of what tools are left to discover the truth if conventional standards of evidence and method are supposed to be inadequate is simply that it is attempts to impose rigid rules on the spectrum of events in the world that is the problem — that policy should be more tentative and prepared to acknowledge (embrace) uncertainty.

Policy, however, is not usually entrusted to those who say they are not sure, who say it is hard to know. In practical life, the prizes all go to those who say they know, who seem to have the answers. As Kuhn puts it too, the nature of expert groups in all areas of life requires would-be entrants to undergo rigorous and rigid preparation that will make it fairly inevitable that the received beliefs exert a deep hold on the mind of each new member of the intellectual community for life.

Indeed, what is considered to be normal science 'is predicated on the assumption that the scientific community knows what the world is like' — and 'responsible' scientists take great pains to defend that assumption. New ideas, new paradigms or theories, have to be suppressed, because they are 'necessarily subversive' of basic commitments. To do otherwise would require the reconstruction of existing assumptions and the re-evaluation of accepted facts. This would be a huge task, possibly dangerous, possibly impractical. Certainly, time consuming. Thus it is to be strongly resisted by sensible members of the established community. 'Novelty emerges only with difficulty, manifested by resistance.'

Nor can the scientists function without a set of beliefs, for a paradigm is essential to scientific enquiry—'no natural history can be interpreted in the absence of at least some implicit body of intertwined theoretical and methodological belief that permits selection, evaluation, and criticism.'

Kuhn also agrees that 'evidence' alone does not decide theories. He notes that philosophers of science have repeatedly demonstrated that more than one theoretical construction can always be placed upon a given collection of data. Even if problems and weaknesses with a theory begin to accumulate, he says, it is easier for the establishment, scientific, religious, political, to either 'modify' the original idea, or more often to suppress the conflicting information, than to abandon their established orthodoxies.

So how, and why, do scientific paradigms ever change? Kuhn links the process to the nature of perceptual (conceptual) change in an individual, where change is resisted at first, but once the 'jump' is made, the old ways of thinking become impossible to return to. It is also, he suggests, analogous to a political revolution.

But the important thing is not so much to passively watch the mud-wrestling of the philosophers—but to ask whether it matters that science moves in 'fits and starts', or the fact that it involves political and social forces, or even that it is simply 'politics by other means'. Don't these facts show, rather, that science is merely (but still importantly) a historical, piecemeal, and fallible process of gaining limited knowledge of the world? And isn't this, rather, a view that sceptics and scientists both ought to defend?

Notes and Key Sources

Chapter 1: Tales of Mice and Men

What are the origins of life? There are two quite different kinds of books on this, the rather dull chemistry ones such as *The Molecular Origins of Life: Assembling Pieces of the Puzzle*, edited by Andre Brack (Cambridge University Press 1998), and ones 'with an agenda' such as *Men of Science-Men of God* by Henry Morris (1982). Probably the best one in this strand is *Origin of Life*, by Aleksander Oparin (Russian original edition 1936, Dover English edition 2003). The sensible sceptic quote the chapter opens with is from an academic paper: 'Molecular Evolution and the Origin of Life', by Sidney W. Fox, Klaus Dose, in *Interdisciplinary Science Reviews*, Volume 13, p. 348.

More colourful alternative and mythological accounts are described by Alan Dundes and contributors in *Sacred Narrative: Readings in the Theory of Myth* (University of California Press 1984).

Chapter 2: Discarding Fossilized Theories

Darwin's magnum opus *Origin of Species* is of course there to be examined, and much more readable than many a contemporary biologist or botanist would manage, but better is to take a good overview of the ideas and issues, of which there are many, such as *The Cambridge Companion to the Origin of Species*, edited by Michael Ruse and Robert Richards (Cambridge 2009), part of a series of books to 'celebrate' some anniversary or other of Darwin. Dangerously one-sided, but otherwise quite entertaining, are the popularizations of evolutionary science in Richard Dawkins' various books including *The God Delusion* (Bantam Press 2006, and other editions). Over several hundred pages, and these largely repeating an earlier book of his, Dawkins explains that natural selection can explain the journey from primordial chemistry to the dazzling variety of life forms we know today. Morality, human consciousness, art, even religion itself, all can be explained by the elegantly simple theory of natural selection.

A good introduction to the range of views is *An Evolving Dialogue: Theological and Scientific Perspectives on Evolution* (Bloomsbury 2001), edited by James B. Miller, and an interesting historical perspective (archived at https://archive.org/details/evolutionoldnewo00unse) is provided by Samuel Butler's classic text, originally published in 1879, *Evolution, Old and New; Or, the theories of Buffon, Dr. Erasmus Darwin, and Lamarck, as compared with that of Charles Darwin.*

Chapter 3: The Brain Doctors

The opening quotation 'for' psychosurgery is by Bengt Jansson, Professor of Psychiatry at Karolinska Institutet, Sweden.

The essay, 'Controversial Psychosurgery Resulted in a Nobel Prize' is available online at the Nobel Prize website: http://www.nobelprize.org/nobel_prizes/medicine/laureates/1949/moniz-article.html

As mentioned in the text, a great introduction to the whole issue is *Amputated Souls: The Psychiatric Assault on Liberty 1935–2011,* by Anthony James (also by Imprint Academic 2013). Lisa Appignanesi's book, also cited, *Mad, Bad and Sad: A History of Women and the Mind Doctors from 1800 to the Present,* was published by Virago in 2008. You can read the debate between the two sides in the UK arguing about the classification of mental illnesses in the article 'Do We Need to Change the Way We Are Thinking About Mental Illness?' in the *Guardian,* archived at http://www.theguardian.com/science/2013/may/12/dsm-5-conspiracy-laughable.

Peter Breggin, cited in the main text, is himself the author of more than twenty books including *Talking Back to Prozac* (1994, with Ginger Breggin), *Medication Madness: The Role of Psychiatric Drugs in Cases of Violence, Suicide and Crime* (2008), and *Psychiatric Drug Withdrawal: A Guide for Prescribers, Therapists, Patients and Their Families* (2013). Another very readable sceptical, indeed downright critical, account of psychiatric practice by Breggin is his book *Toxic Psychiatry: Why Therapy, Empathy and Love Must Replace the Drugs, Electroshock, and Biochemical Theories of the 'New Psychiatry'* (St. Martin's Griffin 1994).

A critique of Breggin and his arguments entitled 'Prozac's Worst Enemy' appeared in *Time Magazine,* 10 October 1994, written by Christine Gorman. You can read the article here: http://www.holysmoke.org/sdhok/dep03.htm.

Chapter 4: Inexplicable Diseases

A thorough and popular, if poorly written, introduction to how the medical industry manipulates its supposedly 'scientific' tests is *Bad Pharma: How Medicine is Broken, and How We Can Fix It* (Fourth Estate 2013) by... Ben Goldacre! Is it a case of poacher turned gamekeeper?

But back to 'Zombie Science: A Sinister Cconsequence of Evaluating Scientific Theories Purely on the Basis of Enlightened Self-Interest' by Bruce Charlton, is a paper in the journal *Medical Hypotheses,* Vol. 71, pp. 327–329, in 2008, which is also very revealing.

The BBC's scary reporting of flu viruses is all archived on its website, at www.bbc.co.uk. The World Health Organization provides a similar service at the addresses below as do the Centers for Disease Control and Prevention. Both sites, by the magic of the internet, are then the source of much scary advice repeated by books, magazines, and other websites.

The alarmed members of the public commenting as 'guests' were on the health website http://www.suite101.com from which some of the 'popular' health advice on flu has also been drawn.

* WHO Declares Swine H1N1 Pandemic for 2009
<http://www.suite101.com/content/who-declares-swine-h1n1-pandemic-for-2009-a113460>
* Epidemic Pandemic of Swine Flu in 2009 Grows
<http://www.suite101.com/content/epidemic-pandemic-swine-flu-2009-expands-grows-a112800>
* How to Protect Your Child from Swine Flu
<http://www.suite101.com/content/how-to-protect-your-child-from-swine-flu-a145576>

The site itself draws on this book: Timbury, M.C., et al. (2002) *Notes on Medical Microbiology,* Churchill Livingstone, New York, 598 pp. The alarming figures about 'Yuppie Flu' are in Haavisto Maija, *Reviving the Broken Marionette: Treatments for CFS/ME and Fibromyalgia.*

The study of the health benefits of drinking tea is called in full: 'Specific Formulation of Camellia sinensis Prevents Cold and Flu Symptoms and Enhances {gamma}{delta} T Cell Function: A Randomized, Double-Blind, Placebo-Controlled Study', authored by Cheryl A. Rowe, PhD, Meri P. Nantz, BA, Jack F. Bukowski, MD, PhD and Susan S. Percival, PhD, and was published in the *Journal of the American College of Nutrition,* Vol. 26, No. 5, pp. 445–452 (2007).

Comparisons with simple things like providing clean water are always useful, if much ignored. D. Narayan, *The Contribution of People's Participation, Evidence from 121 Rural Water Supply Projects, Environmentally Sustainable Developmental Occasional Paper Series No. 1* (Washington DC, The 12. World Bank 1995), p. 59.

World Bank, *World Development Report 1994: Infrastructure for Development* (Oxford University Press 1995), p. 11 and p. 83; and Kofi Annan, *Progress Made in Providing Safe Water Supply and Sanitation for all*

During the 1990s. Report of the Secretary-General. Economic and Social Council, Commission on Sustainable Development, 8th session.

Chapter 5: Inexplicable Cures

The Stanford quote at the beginning of the chapter is from *Science and Pseudo-Science*, by Sven Ove Hansson, with the webpage published Wed 3 September 2008. My main source on Hahnemann is: *The History of Homeopathy: Its Origin; Its Conflicts*, by Wilhelm Ameke. Wilhelm Ameke (1847–1886) was himself an orthodox doctor who converted to homeopathy to become one of the great historians of Samuel Hahnemann and of homeopathy. He died, aged just 39, of tuberculosis.

The statistics on how negative journal authors are about homeopathy comes from 'A Systematic Review of How Homeopathy is Represented in Conventional and CAM Peer Reviewed Journals', Timothy Caulfield and Suzanne DeBow, *BMC Complementary and Alternative Medicine*, 2005, 5, p. 12.

Robert van der Bosch describes the cancer conspiracy in the book *The Pesticide Conspiracy* (University of California Press 1989). The issue is also raised in Fred Harding's more recent book, *Cancer: Cause – Prevention – Cure*. Brian Josephson is commenting on homeopathy in the article 'Molecule Memories', *New Scientist*, 1 November 1997. The Belfast study of ultra-dilute remedies was published in the journal *Inflammation Research*, vol. 53, p. 151.

The BBC health correspondent, Pallah Ghosh, describes the investigation in an article called 'Homeopathic Practices Risk Lives', published on the BBC website on 13 July 2006. And one trifling detail that has vexed researchers into not so much alternative medicine as prominent alternative medicine critics is that although Ben Goldacre has a degree that means he is certainly qualified as a doctor, and although, as mentioned in the chapter, his book states firmly that he 'works full-time as a doctor for the NHS', the UK registry of doctors in the National Health Service (as of 2014) records that he has thus far not registered to practise his healing arts in either hospitals or doctor's surgeries. In 2013 he was still promoting his books on his website, www.badscience.net, with hints at his 'life and death' responsibilities thus:

> 'Although we work with life and death, doctors, academics, pharmacists, patient groups, regulators and the rest of us in medicine are the same as any other set of workers. We can all easily fall into a rhythm of "getting by", responding reflexly to proximate incentives: getting the next grant, getting the next academic publication, getting through the next clinic, processing the paperwork, and keeping our heads above the waterline.'

Finally, for the 'orthodox view' of the benefits of homeopathy, I consulted a modest but clear and careful booklet, part of a series on 'alternative health' by Sarah Richardson called 'A Guide to Homeopathy', published originally in 1988.

Chapter 6: Physics' Guilty Secrets

Newton's *Opticks* (1704) postulated an 'Aethereal Medium' transmitting vibrations faster than light. Newton believed that these vibrations were related to heat radiation saying that heat resulted 'when a Ray of Light falls upon the Surface of any pellucid Body'. He added: 'I do not know what this Aether is', but that if it consists of particles then they must be 'exceedingly smaller than those of Air, or even than those of Light.' Similarly, Paul Drude wrote in 1900: 'The conception of an aether absolutely at rest is the most simple and the most natural — at least if the aether is conceived to be not a substance but merely space endowed with certain physical properties' (P. Drude, *The Theory of Optics*, trans. by C.R. Mann and R.A. Millikan, Dover, New York, 1959, p. 457).

The classic work on the subject, by E.T. Whittaker, mentioned in the text, puts it this way:

> 'It is thus erroneous to regard the heavenly bodies as isolated in vacant space; around and between them is an incessant conveyance and transformation of energy. To the vehicle of this activity the name aether has been given. The aether is the solitary tenant of the universe, save for that infinitesimal fraction of space which is occupied by ordinary matter. Hence arises a problem which has long engaged attention, and is not yet completely solved: What relation subsists between the medium which fills the interstellar void and the condensations of matter that are scattered through it?' (*Theories of Aether and Electricity*, E.T. Whittaker, Longmans, Green and Co., Dublin, 1910, p. 1)

Lastly, details of that paper in full: 'Aether Theories: A Physics Fairy Tale Re-told', by K.M Hajek and T.A. Nieminen, in the *Australian Institute of Physics 17th National Congress*, 2006 — Brisbane, paper 599, p. 4. Einstein admits all about the aether in a paper called in German, 'Grundgedanken und Methoden der Relativitätstheorie, in ihrer Entwicklung dargestellt', published in 1920.

An interesting article about Crystal quanta and space too by Dean D'Souza appears in the *Essentials of Philosophy and Ethics* (edited by me) and published by Hodder in 2006.

Chapter 7: Black Holes, God Particles, and Bombast

'The more you see how strangely Nature behaves, the harder it is to make a model that explains how even the simplest phenomena actually work. So theoretical physics has given up on that.' From: *QED: The Strange Theory of Light and Matter*, by Richard Feynman (1985). This book is an adaptation for the general reader of four lectures on quantum electrodynamics.

'Science is the belief in the ignorance of experts' is from Feynman's talk 'What is Science?', presented at the fifteenth annual meeting of the National Science Teachers Association, in New York City (1966), and published in *The Physics Teacher*, volume 7, issue 6 (1969), pp. 313–320.

'Others have suggested that the speed of light used to be even higher, but has since slowed down.' For example, John Moffat, a Canadian scientist, in 1992, and Joao Magueijo, a professor at Imperial College, London, in his book *Faster than the Speed of Light* (2003).

'If the speed of light is the least bit affected by the speed of the light source, then my whole theory of relativity and theory of gravity is false.' This is one of a series of *mea culpa* quotations attributed to Einstein. In similar spirit he also said: 'I consider it quite possible that physics cannot be based on the field concept, i.e., on continuous structures. In that case, nothing remains of my entire castle in the air, gravitation theory included, [and of] the rest of modern physics.'

In his 1905 paper on Special Relativity (short, technical of course, but still interesting and easily found online), and during later years, Albert Einstein worked out some of the main physical consequences of his theory, like length contraction, time dilation, and the much celebrated formula $E = mc^2$. A century on, these effects have all been experimentally confirmed and it would appear that Special Relativity's success is a confirmation that Einstein's assumption of constancy of the one-way velocity of light is an actual property of nature—a true physical quantity. However, a group of philosophers of science, known as the conventionalists, have argued this to be a misconception. A paper published in 1970 by John Winnie, from the University of Hawaii, has been able to reformulate Special Relativity without the one-way velocity of light assumption, calculating effects such as length contraction and time dilation experimentally identical to what Einstein derived in 1905.

Gerald Holton, a professor of physics at Harvard, not only confirmed Einstein's judgment of the non-revolutionary character of the theory, but described it as an example of the general rule that a 'so-called scientific "revolution" turns out to be at bottom an effort to return to a classical purity.' See G. Holton, 'On the Origins of the Special Theory of Relativity', *American Journal of Physics*, 28 (1960): pp. 627–636; reprinted in G. Holton, *Thematic Origins of Scientific Thought: Kepler to Einstein* (Harvard University Press 1973), pp. 165–183.

William Newman, Professor in the History and Philosophy of Science at Indiana University, was commenting in an opinion article entitled 'Isaac Newton: The Last Magician' for *Voice Magazine*. The whole article is available online at: http://voice.unimelb.edu.au/volume-10/number-9/isaac-newton-last-magician.

How original was Einstein anyway? Muneeb Faiq, a contemporary scientist with diverse philosophical interests, has pointed out that scientists such as J.B. Stallo, Streintz, Everett, and L. Lange had already detailed much of the theory of relativity before Einstein. He says that what the world applauded as the 'Principle of Relativity' was a restatement by Einstein (with his own mathematical contribution of work including: L. Lange, 'Das Inertial system vor dem forum der Naturforschung: Kritischus und Antikritichus', *Philosophischie Studien*, Vol. 20 (1902), p. 18; J.B. Stallo, *Die Begriffe und theorieen der modernen physic*, Johann Ambrosius Barth, Leipzig, (1901), pp. 205, 331).

It is notable that Einstein's papers on Special Relativity, $E = mc^2$, Brownian Motion, and the Einstein–Podolsky–Rosen paradox contained no citations at all! However, his paper on General Relativity contained some.

Chapter 8: Spooky Coincidences and Amazing Insights

For an exciting but unkind overview of esoteric see *Memoirs of Extraordinary Popular Delusions and the Madness of Crowds*, by Charles Mackay. Similarly 'Objections to Astrology: A Statement by 186 Leading Scientists' appeared in *The Humanist* magazine for September/October 1975. Maybe don't bother to follow up Lawrence E. Jerome, *Astrology Disproved*, New York: Prometheus Books, 1977.

The detail about Newton discovering maths in astrology is given in an essay by J. O'Connor and E.F. Robertson. See http://www.history.mcs.st-and.ac.uk/Biographies/Newton.html.

On the other side are Jung, C.G. (1958) 'An Astrological Experiment', in *The Secret Life*, CW18 (London: RKP 1977 edition); and Koestler, A. (1972) *The Roots of Coincidence* (London: Picador). Michael Shallis makes the relevant distinction between descriptive and explanatory science in relation to synchronicity in *On Time* (1982, Burnett Books). Two other books on synchronicty are Progoff, I. (1973) *Jung, Synchronicty and Human Destiny* (New York: Julian Press) and Von Franz, M. (1980) *On Divination and Synchronicity* (Toronto: Inner City Books). A good overview, drawing on these kinds of sources, is by Maggie Hyde, sometime Director of the Company of Astrologers. Her book is called *Jung and Astrology* (1992).

Chapter 9: Bubbles, Black Swans, and Banking Disasters

What to read on economics? The subject has that reputation for being dismal, and books on it bear that out. Articles in newspapers are often better—but then they seem too superficial to be worth bothering with—reflecting only the week's events, not explaining the long-term reality. But can anyone do that? Inevitably, such a quest returns the seeker of economic wisdom to just one or two books, Adam Smith's venerable *Wealth of Nations,* and J.K. Galbraith's *The Affluent Society* (1958) are undoubted classics. A recent popularization with a nice clear style and a passionate stance is *Economics: The User's Guide* by Ha-Joon Chang (2014).

'The policies affecting liquidity created a situation like a dam...' From: 'The Current Financial Crisis: Causes and Policy Issues', Adrian Blundell-Wignall, Paul Atkinson and Se Hoon Lee.

'By 2007, the combined assets of the Quants specialising in US stocks were comfortably over $1.2 trillion'. See, for example, *The New York Times*: '"Quant" Funds Try to Resurrect Computer Investing', www.nytimes.com/2010/08/20/business/20quant.html.

'When it comes to investing, our incorrigible search for patterns leads us to assume that order exists where if often doesn't.' Quoted in *The Guru Investor: How to Beat the Market Using History's Best Investment* by John P. Reese, Jack M. Forehand (Wiley 2009).

In economics, *Homo economicus*, the term being a play upon *Homo sapiens*, for thinking human, is the notion underlying many economic theories that humans are rational and self-interested actors who have the ability to make judgments toward their subjectively defined ends.

'How did Goldman Sachs know to disinvest in BP...?' According to regulatory filings, reported by RawStory.com, has found that Goldman Sachs sold 4,680,822 shares of BP in the first quarter of 2010. Goldman's sales were the largest of any firm during that time. Goldman would have pocketed slightly more than $266 million if their holdings were sold at the average price of BP's stock during the quarter.

'It's funny, but when quants do well, they all call themselves brilliant, but when things don't go well, they whine and call it an "anomalous market"' —as reported in the *New York Times*, 'Shrinking "Quant" Funds Struggle to Revive Boom' by Julie Cresswell, 19 August 2010.

'We love the tangible, the confirmation, the palpable...' *The Black Swan,* byNassim Nicholas Taleb, Random House, 2010 (Second edition).

Antonio Borges, former bigwig at Goldman Sachs, now poacher turned gamekeeper, was quoted in 'Crisis Is Over, but Where's the Fix?', by Floyd Norris for the *New York Times*, 10 March 2011.

The figures on the collapse of value of Quant funds are according to eInvestment Alliance, a research firm, while Floyd Norris is opining in the *IHT*, 11 March 2011, 'Wasting a Chance to Fix Finance'.

Chapter 10: Climate Science and the Profits of Doom

The core of this chapter is based on research for two long articles I wrote for the *Times Higher Education* (London), notably: 'Beyond Debate' (9 December 2009) and 'Profits of Doom' (22 July 2010), both of which are available online, albeit without the very nice pictures in the original magazine. The original articles contain more detail – but there is also a vigorous debate and some additional ideas in the numerous comments that follow the articles. Because climate science is a topic that many people do take a view on.

Behind that are numerous specialist papers, such as the IPCC reports, especially the Summaries for Policy Makers, papers for various obscure scientific journals such as 'General Circulation Modelling of Holocene, that is, the Current Climate Period Climate Variability' by 'warmists' Gavin Schmidt, Drew Shindell, Ron Miller, Michael Mann and David Rind in *Quaternary Science Review*, 23 (2004); and 'On the Determination of Climate Feedbacks from ERBE Data' by (fence-sitters) Richard S. Lindzen and Yong-Sang Choi; and not to forget numerous contrarian reports by climate sceptics, often restricted to the internet. Neither body of information has survived the test of time. If you want a good objective assessment of 'climate science', however, a book that came out 'post-Copenhagen', as it were, *Climate: The Counter Consensus*, by Robert Carter (Stacey International 2010), says all you need to know.

The section on the climate researcher Hubert Lamb draws on research by Bernie Lewin, published in various places online, but collected in large part together for a special essay for the Global Warming Policy Foundation entitled 'Hubert Lamb and the Trans-formation of Climate Science' available on its website at www.gwpf. org. The essay draws in turn on these sources. The quotes on 'powerful interests at work' and on 'fashions in scientific work' are from H.H. Lamb, *Weather, Climate & Human Affairs: A Book of Essays and Other Papers*, London: Routledge, 1988. Lamb looks at climate science and worries that '...the prospects of global warming are now spoken of on every side and are treated by many, including people whose decisions affect millions, as if the more alarming forecasts were already *established* as fact' in *Climate, History, and the Modern World* (2nd ed., London; New York: Routledge, 1995).

Hansen's research predicting a global warming of 'almost unprece-dented magnitude' was reported in W. Sullivan, 'Study Finds Warming Trend That Could Raise Sea Levels', *The New York Times*, p. 1, 22

August 1981. The 'precarious and threatening situation' in climatology is described in the *World Meteorological Organization Bulletin*, vol. 43, no. 2, April 1994.

The scholarly work, *International Environmental Policy*, by Aynsley Kellow and Sonja Boehmer-Christiansen (Edward Elgar 2002), provides an authoritative insight into the political forces driving climate research. A later work (2010), *The International Politics of Climate Change*, is a good resource for libraries — it costs hundreds of zlotys! A pity as it has a good look at what it calls 'the Place of Science'.

John Holdren's warning is in Paul Ehrlich's (one of global warming theory's most extravagant doomsayers) book, *The Machinery of Nature* (Simon & Schuster, 1986, p. 274). Finally, Bruce Yandle, an economist at Clemson University in the US, coined the phrase 'Baptist and Bootlegger coalitions' in an article in *Regulation Magazine* in 1983.

Chapter 11: The Risk Factor

This is a controversial area but here are a few sources and reports to start with.

Helen Caldicott was opposing nuclear power in a debate called 'Prescription_for_Survival' on 30 March 2011. You can see the debate at: http://www.democracynow.org.

Sources on radiation effects include the Reuters report 'Ocean Currents Will Disperse the Radiation' by Yoko Kubota, 'Engineers Toil to Pump Out Japan Plant', 26 March 2011, http://www.reuters.com/article/2011/03/26/us-japan-quake-idUSTRE72A0SS20110326.

The hidden story of nuclear's contract workers is revealed by a French article at *L'Humanité*: 'Pour travailler à Fukushima, il faut être prêt à mourir'. Interview by Anne Roy, translated 7 April 2011, by Henry Crapo and reviewed by Bill Scoble.

A summary by Prof. Paul Jobin, 'Dying for TEPCO? Fukushima's Nuclear Contract Workers', is at http://www.globalresearch.ca/index.php?context=va&aid=24543. And see also the *New York Times* feature by Hiroko Tabuchi, 'Japanese Workers Braved Radiation for a Temp Job', 9 April 2011, http://www.nytimes.com/2011/04/10/world/asia/10workers.html.

Debora McKenzie reported that 'Fukushima Radioactive Fallout Nears Chernobyl Levels', in the 24 March 2011 issue of the *New Scientist*, http://www.newscientist.com/article/dn20285-fukushima-radioactive-fallout-nears-chernobyl-levels.html.

The report, *Chernobyl: Consequences of the Catastrophe for People and the Environment*, vol. 1181 (New York: Annals of the New York Academy of Sciences, 2009) is by Alexey Yablokov, Vassily Nesterenko, and Alexey Nesterenko.

What about the children in the orphanages? Did they exist or not? The question is perhaps settled by shocking photographs, taken by Paul Fusco, of an orphanage in Minsk filled with children deformed by Chernobyl which can be found at http://inmotion.magnumphotos. com/essay/chernobyl.

For a bit of light reading but also many more atomic facts, the story of the early atomic tests in the Pacific is in my political travel guide, *No Holiday: 80 Places You Don't Want to Visit* (New York: Disinformation Press 2006).

Medical studies include Emilia A. Diomina, 'Radiation Epidemiological Studies in a Group of Liquidators of the Chernobyl Accident Consequences', R.E. Kavetsky Institute of Experimental Pathology, Oncology and Radiobiology of National Academy of Sciences of Ukraine, Kiev, 2006.

The *Guardian* website also carries a 10 January 2010 article, 'Chernobyl Nuclear Accident: Figures for Deaths and Cancers Still in Dispute', by John Vidal, its environment editor, adopting a position entirely contrary to that of the paper's environment columnist George Monbiot, along with another responding to the WHO analysis titled, 'UN Accused of Ignoring 500,000 Chernobyl Deaths', 25 March 2006. Interestingly, for a journalist who preaches the importance of both consulting and giving your sources, a recent article where George Monbiot declared his new found love for nuclear power offers as one of his sources 'xkcd.com', which for those of you who do not frequent it regularly is a 'web comic', the author of which states firmly: 'I'm not an expert in radiation and I'm sure I've got a lot of mistakes in here.' Bravo for xkcd science!

Monbiot would have been pleased to see Dr Windridge's broadside *for* nuclear appear in the *Guardian*, 'Fear of Nuclear Power Is Out of All Proportion to the Actual Risks', 4 April 2011, http://www.guardian.co. uk/science/blog/2011/apr/04/fear-nuclear-power-fukushima-risks.

'The WHO expert group placed particular emphasis on scientific quality', promised WHO Fact Sheet No. 303, 'Health Effects of the Chernobyl Accident: An Overview', published in April 2006. The influential World Health Organization report is summarized in 'Health Effects of the Chernobyl Accident: An Overview', Fact Sheet No. 303, *WHO*, April 2006.

Chapter 12: African Art or High Street Kitsch?

Regarding the Matisse sketch, the answer is that the faithless forgery is on the left, and the original is on the right. Apparently.

The history and philosophy of art and aesthetics is a long and, well, flowery one. For example, the first study of the value of the sublime is the treatise ascribed to Longinus: *On the Sublime (Peri hupsous)*. This is a document thought to have been written as early as the first century AD, but first published in 1554. Saint Augustine, Immanuel Kant, and Schopenhauer and Nietzsche all had their two penn'orth too. For this last, see for example, *Human, All Too Human* in *Volume Two: Assorted Opinions and Maxims* (Nietzsche, F., *Human, All Too Human: A Book for Free Spirits*, trans. R.J. Hollingdale, Cambridge University Press, Reprint, 2005).

Experts are attached to Alexander Pope's *An Essay on Criticism* in 1711 whose insight has been summarized obscurely in two lines as follows: 'Whose own Example strengthens all his Laws, and Is himself the great Sublime he draws' (lines 679–680). A more recent and indeed more interesting account though is that of Nancy Etcoff, in the pithily entitled *Survival of the Prettiest* (1999). Here she argues that 'beauty' is more than in the eye of the beholder — it is a cross-cultural reality.

Afterword: Paradigm Shifts

John Hogan's talk is recorded in the magazine *EDGE* 16 for 6 May 1997, and his book was published by Broadway Books in 1997.

Karl Popper on 'Falsification' is in his classic book, *The Poverty of Historicism*, pp. 133–4, 1957, second edition, Routledge, 1960. The 'standard' texts on philosophy of science include in pride of place *The Logic of Scientific Discovery* (translation of *Logik der Forschung*, 1935), Hutchinson, London, 1959; and *The Structure of Scientific Revolutions* (1962) by Thomas S. Kuhn. A non-standard but more entertaining text is Paul Feyerabend's *Three Dialogues on Knowledge*, published by Blackwell in 1991. This is actually written as little playlets, in the manner of the master (Plato) himself, but unlike many other efforts to use the format, Feyerabend is quite good at it. Otherwise Feyerabend's *Against Method* remains the champion of anarchist theories of knowledge — to use the category he chooses for himself.

About the Author

Martin Cohen is a well-established author specializing in popular books in philosophy, social science, and politics. He is best known for his two introductions to philosophy, *101 Philosophy Problems* (Routledge 1999, 2001, 2007) and *101 Ethical Dilemmas* (Routledge 2002/2007) which, despite being originally aimed at the academic market, between them have sold over 250,000 copies and been translated into 20 different languages. He also published a popular overview of political philosophy for Pluto Press (*Political Philosophy from Plato to Mao*, 2001, 2008) and an 'anti-history' of great philosophers, *Philosophical Tales* (2008) for Blackwell. A book on thought experiments was well-reviewed despite being entitled (confusingly perhaps!) *Wittgenstein's Beetle* (2004), and other more academic books include a mini-book on Adam Smith; and a reference guide (a kind of mini-encyclopaedia) to philosophy and ethics for Hodder Academic.

His most recent projects include *Critical Thinking Skills for Dummies* (Wiley 2015) and the UK edition of *Philosophy for Dummies* (Wiley, June 2010); *Mind Games: 31 Days to Rediscover Your Brain* (Blackwell, July 2010); *The Doomsday Machine: The High Price of the World's Most Dangerous Fuel* (Palgrave 2012), a book which the *New York Times* described as a 'polemic against the atom' and discussed for its unusual take on climate change politics, and a popular series of books on the theme 'How to Live' (Media Studies Unit 2013 and 2014) featuring the wise (and not-so-wise) advice of the Ancients.

Martin Cohen has a degree in philosophy and social science from Sussex University, a post-graduate diploma in information technology from Hull University, a PGCE from Keele, and a PhD in philosophy and education (for a thesis entitled: 'What is the Educational Value of Information Technology?') from Exeter University. He now writes full-time, but in the past has taught philosophy and social science at a number of universities in the UK and Australia, and was involved in a research project at Leeds University under George MacDonald Ross exploring ways to shift philosophy teaching away from the mere study of philosophical facts and toward a view of philosophy as an activity.

Martin is also the editor of *The Philosopher*, a venerable journal founded in 1923 with a mission to bring 'philosophy to all', and now published (entirely open access) online at www.the-philosopher.co.uk. *The Philosopher* counts some of the best known names in twentieth-century philosophy amongst its contributors. His editorial strategy is to allow as wide a range of ideas as possible a forum, and often prints papers by non-specialists with unusual and original ideas. He is currently based in Aquitaine, France, where he lives with his French artist wife, Judith, and young son, Milo, but travels often to the US and UK.

Index

www.ingramcontent.com/pod-product-compliance
Lightning Source LLC
Chambersburg PA
CBHW061731270326
41928CB00011B/2187